THE SUNDAY

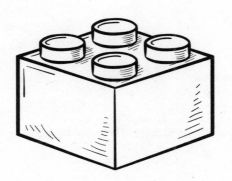

TEASERS

BOOK 2

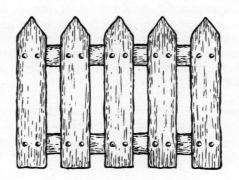

EDITED BY JOHN OWEN

Published in 2023 by Times Books
HarperCollinsPublishers
Westerhill Road
Bishopbriggs
Glasgow, G64 2QT

HarperCollins Publishers
Macken House
39/40 Mayor Street Upper
Dublin 1, D01 C9W8
Ireland

10 9 8 7 6 5 4 3 2 1

The Sunday Times is a registered trademark of Times Newspapers Ltd

ISBN 978-0-00-861796-7

Typeset by Davidson Publishing Solutions, Glasgow

Printed and bound by CPI Group (UK) Ltd, Croydon CR0 4YY

Images used under license from Shutterstock.com

If you would like to comment on any aspect of this book, please contact us at
the above address or online via email: puzzles@harpercollins.co.uk

This book is produced from independently certified FSC™ paper
to ensure responsible forest management.

For more information visit: www.harpercollins.co.uk/green

INTRODUCTION

Welcome to our second collection of *Sunday Times* Teasers. Here we bring together another 100 mind-bending lateral-thinking puzzles from the series, which has been tantalising mathematically inclined readers for more than 60 years.

As well as appearing in the print paper, the weekly Teasers can be accessed by *Times* subscribers through a link on the Sunday puzzles page at thetimes.co.uk.

The Teasers take the form of a paragraph or two of text, sometimes accompanied by a diagram, and require varying degrees of mathematical or logical reasoning to reach the answer.

In this collection we also provide full solutions to each of the puzzles. These go beyond the brief answer given in the newspaper, setting out exactly how to get to the answer, which I hope will help readers to understand the puzzles and to improve their solving skills.

The Teasers in this book were originally published between December 2020 and November 2022, and have all been edited by the brilliant John Owen. I am very grateful for the work John puts into each puzzle before it appears in the paper, as well as for the extra effort he has put into producing this book.

John contributes puzzles to the series (you can see some of his work on the pages of this collection), complementing those submitted by many other regular and occasional setters.

The ingenuity of our setters is extremely impressive, and I would like to thank them all for the entertainment they continue to provide to readers.

All Teaser submissions are unsolicited and anyone with a good idea is welcome to send in a puzzle. If you have been inspired by the challenges in this book and would like to write a Teaser puzzle for publication in *The Sunday Times*, please email puzzle.feedback@sunday-times.co.uk for more information.

Mick Hodgkin

Puzzles Editor, The Times and The Sunday Times

PUZZLES

MOVING DIGIT

Andrew Skidmore

Jonny has opened a new bank account and has set up a telephone PIN. His sort code is comprised of the set of three two-figure numbers with the smallest sum that give his PIN as their product. He was surprised to find that the PIN was also the result of dividing his eight-figure account number by one of the three two-figure numbers in the sort code.

The PIN has an unusual feature that Jonny describes as a moving digit. If the number is divided by its first digit then the number that results has the same digits in the same order, except that first digit is now at the end.

The account number does not contain the digit that moved.

What is the account number?

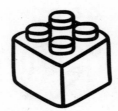

THE BEST GAME PLAYED IN 1966

Stephen Hogg

For Christmas 1966 I got 200 Montini building blocks; a World Cup Subbuteo set; and a Johnny Seven multi-gun. I built a battleship on the 'Wembley pitch' using every block, then launched seven missiles at it from the gun. The best game ever!

Each missile blasted a different prime number of blocks off the 'pitch' (fewer than remained). After each shot, in order, the number of blocks left on the 'pitch' was: a prime; a square; a cube; a square greater than 1 times a prime; a cube greater than 1 times a prime; none of the aforementioned; and a prime.

The above would still be valid if the numbers blasted off by the sixth and seventh shots were swapped.

How many blocks remained on the 'pitch' after the seventh shot?

3

A QUESTION OF SCALE

Colin Vout

The modernist music of Skaredahora eschewed traditional scales; instead he built scales up from strict mathematical rules.

The familiar major scale uses 7 notes chosen from the 12 pitches forming an octave. The notes are in (1) or out (0) of the scale in the pattern 101011010101, which then repeats. Six of these notes have another note exactly 7 steps above (maybe in the next repeat).

He wanted a different scale using 6 notes from the 12 pitches, with exactly two notes having another note 1 above, and one having another 5 above. Some notes could be involved in these pairings more than once.

His favourite scale was the one satisfying these rules that came first numerically when written out with 0s & 1s, starting with a 1.

What was Skaredahora's favourite scale?

4

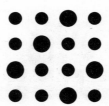

AN ODD SELECTION

Victor Bryant

I wrote an odd digit in each of the sixteen cells of a four-by-four grid, with no repeated digit in any row or column, and with each odd digit appearing three or more times overall. Then I could read four four-figure numbers across the grid and four four-figure numbers down. I calculated the average of the four across numbers and the larger average of the four down numbers. Each was a whole number consisting entirely of odd digits, and each used an odd number of different odd digits.

What were those two averages?

5

BUILDING BLOCKS

Colin Vout

The kids had used all the blocks each of them owned (fewer than 100 each) to build triangular peaks – one block on the top row, two on the next row, and so on.

"My red one's nicer!" "My blue one's taller!"

"Why don't you both work together to make one bigger still?" I said. I could see they could use all these blocks to make another triangle.

This kept them quiet until I heard, "I bet Dad could buy some yellow blocks to build a triangle bigger than either of ours was, or a red and yellow triangle, or a yellow and blue triangle, with no blocks ever left over."

This kept me quiet, until I worked out that I could.

How many red, blue and yellow blocks would there be?

6

LET TEL PLAY BEDMAS HOLD 'EM!

Stephen Hogg

Awaiting guests on poker night, Tel placed (using only clubs, face-up, in order left-to-right) the Ace (=1) to 9 (representing numerals), then interspersed the Ten, Jack, Queen and King (representing -, +, x and ÷ respectively) in some order, but none together, among these.

This 'arithmetic expression' included a value over 1000 and more even than odd numbers. Applying *BEDMAS* rules, as follows, gave a whole-number answer. "No *B*rackets or pow*E*rs, so traverse the expression, left-to-right, doing each *D*ivision or *M*ultiplication, as encountered, then, again left-to-right, doing each *A*ddition or *S*ubtraction, as encountered."

Tel's auntie switched the King and Queen. A higher whole-number answer arose. Tel displayed the higher answer as a five-card 'flush' in spades (the Jack for + followed by four numeral cards).

Give each answer.

7

SQUARE PHONE NUMBERS

Danny Roth

George and Martha run a business in which there are 22 departments. Each department has a perfect-square three-digit extension number, i.e. everything from 100 (10^2) to 961 (31^2), and all five of their daughters are employees in separate departments.

Andrea, Bertha, Caroline, Dorothy and Elizabeth have extensions in numerical order, with Andrea having the lowest number and Elizabeth the highest. George commented that, if you look at those of Andrea, Bertha and Dorothy, all nine positive digits appear once. Martha added that the total of the five extensions was also a perfect square.

What is Caroline's extension?

8

SOME PERMUTATIONS

Howard Williams

I gave Robbie three different, single-digit, positive whole numbers and asked him to add up all the different three-digit permutations he could make from them. As a check for him, I said that there should be three threes in his total. I then added two more digits to the number to make it five digits long, all being different, and asked Robbie's mother to add up all the possible five-digit permutations of these digits. Again, as a check, I told her that the total should include five sixes.

Given the above, the product of the five numbers was as small as possible.

What, in ascending order, are the five numbers?

9

NERO SCRAG FROM CARREGNOS

Stephen Hogg

Our holiday rep, Nero, explained that in Carregnos an eight-digit total of car registrations results from combinations of three Greek capital letters after four numerals (e.g. 1234 ΩΘΦ), because some letters of the 24-letter alphabet and some numerals (including zero) are not permitted.

For his own 'cherished' registration the number tetrad is the rank order of the letter triad within a list of all permitted letter triads ordered alphabetically. Furthermore, all permitted numeral tetrads can form such 'cherished' registrations, but fewer than half of the permitted letter triads can.

Nero asked me to guess the numbers of permitted letters and numerals. He told me that I was right and wrong respectively, but then I deduced the permitted numerals.

List these numerals in ascending order.

10

PLANTATION

Andrew Skidmore

Jed farms a large, flat, square area of land. He has planted trees at the corners of the plot and all the way round the perimeter; they are an equal whole number of yards apart.

The number of trees is in fact equal to the total number of acres (1 acre is 4840 square yards).

If I told you an even digit in the number of trees you should be able to work out how far apart the trees are.

How many yards are there between adjacent trees?

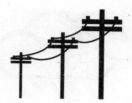

POWER STRUGGLE

Victor Bryant

Given any number, one can calculate how close it is to a perfect square or how close it is to a power of 2. For example, the number 7 is twice as far from its nearest perfect square as it is from its nearest power of 2. On the other hand, the number 40 is twice as far from its nearest power of 2 as it is from its nearest square.

I have quite easily found a larger number (odd and less than a million!) for which one of these distances is twice the other.

What is my number?

12

SHIPPING FORECAST

Colin Vout

"Here is the shipping forecast for the regions surrounding our island.

"First, the coastal regions. Hegroom: E 6, rough, drizzle, moderate. Forkpoynt: E 7, rough, rain, good. Angler: NE 7, rough, drizzle, moderate. Dace: NE 7, smooth, drizzle, moderate.

"Now, the offshore regions. Back: E gale 8, rough, rain, moderate. Greigh: E 7, smooth, drizzle, poor. Intarsia: SE 7, rough, drizzle, moderate. Catter: E gale 8, high, drizzle, moderate. Eighties: E7, rough, fair, good."

In this forecast, no element jumps from one extreme to another between adjacent regions. The extremes are: wind direction, SE & NE; wind strength, 6 & gale 8; sea state, smooth & high; weather, fair & rain; visibility, good & poor.

Each region adjoins four others (meeting at just one point doesn't count).

Which of the regions does Angler touch?

13

PORCUS FACIT GRIPHUS

Stephen Hogg

'Argent bend sinister abased sable in dexter chief a hog enraged proper' blazons our shield (shaped as a square atop a semi-circle, with a 45° diagonal black band meeting the top corner). We've three shields. For the first, in centimetres, the top width (L) is an odd perfect cube and the vertical edge height of the band (v) is

an odd two-figure product of two different primes. The others have, in inches, whole-number L (under two feet) and v values (all different). For each shield, the two white zones have almost identical areas. All three v/L values, in percent, round to the same prime number.

Give the shortest top width, in inches.

14

ENDLESS NUMBER

Danny Roth

George and Martha have a telephone number consisting of nine digits; there is no zero and the others appear once each. The total of the digits is obviously 45, so that the number is divisible by nine. Martha noticed that, if she removed the last (i.e. the least significant) digit, an eight-digit number would remain, divisible by eight. George added that you could continue this process, removing the least significant digit each time to be left with an n-digit number divisible by n right down to the end.

What is their telephone number?

15

DISCS A GO-GO

Howard Williams

My kitchen floor is tiled with identically-sized equilateral triangle tiles while the floor of the bathroom is tiled with identically-sized regular hexagon tiles, the tiles being less than 1m across. In both cases the gaps between tiles are negligible. After much experimenting I found that a circular disc dropped at random onto either the kitchen or bathroom floor had exactly the same (non-zero) chance of landing on just one tile.

The length of each side of the triangular tiles and the length of each side of the hexagon tiles are both even triangular numbers of mm (i.e. of the form 1+2+3+...).

What are the lengths of the sides of the triangular and hexagonal tiles?

16

PROCEED TO CHECKOUT

Colin Vout

The dartboard at the Trial and Arrow pub is rather different from the standard one: there are only 3 sectors, each with a positive whole number label, and no central bullseye-scoring region. There are still double and treble rings: for instance, if the sector label is 3, a dart in that sector can score 3, 6 or 9.

As usual, scores are counted down from a starting value, the idea being to finish ('check out') by reaching exactly zero. Players take turns throwing three darts, or fewer if they check out before that. Unusually, the checkout doesn't have to finish on a double.

The lowest impossible checkout is the smallest value that can't be achieved in one turn; on this board that value is 36.

What are the sector labels?

17

ROSE GARDEN

Andrew Skidmore

The six rose bushes in my garden lie on a circle. When they were very small, I measured the six internal angles of the hexagon that the bushes form. These were three-digit whole numbers of degrees. In a list of them, of the ten digits from 0 to 9, only one digit is used more than once and only one digit is not used at all. Further examination of the list reveals that it contains a perfect power and two prime numbers.

In degrees, what were the smallest and largest angles?

18

CUT FOR PARTNERS

Danny Roth

George and Martha are playing bridge with an invited married couple. Before play starts, the players have to cut for partners. Each player draws a card from a standard pack and those drawing the two highest-ranking cards play together against the other two. For this purpose, the rank order is ♠A, ♥A, ♦A, ♣A, ♠K, ♥K etc. down to ♦3, ♣3, ♠2, ♥2, ♦2, ♣2 (the lowest).

George drew the first card, then Martha drew a lower-ranking card. "That is interesting!" commented the male visitor to his wife. "The probability that we shall play together is now P. Had Martha drawn the ♥7 instead of her actual card, that chance would have been reduced to P/2, and had she drawn the ♥6, the chance would have been reduced to P/3.

Which cards did George and Martha draw?

19

Ω

TOTAL RESISTANCE

Peter Good

A physics teacher taught the class that resistors connected in series have a total resistance that is the sum of their resistances while resistors connected in parallel have a total resistance that is the reciprocal of the sum of their reciprocal resistances, as shown in the diagrams.

$$R = r_1 + r_2$$ $$1/R = 1/r_1 + 1/r_2$$

Each pupil was told to take five 35-ohm resistors and combine all five into a network, using only series and parallel connections. Each pupil then had to calculate theoretically and check experimentally the resistance of his or her network. Every network had a different resistance and the number of different resistances was the maximum possible. The total sum of these resistances was a whole number.

How many pupils were there in the class and what was the sum of the resistances?

20

NINE BLOCKS TO THE DINER

Colin Vout

"Sure, I can help. From this hotel it's nine blocks to the diner; from there it's five blocks to the library, and then six blocks back here. I guess instead you could come back from the diner to the museum – that's four blocks – and then seven blocks back here. Or three blocks from the diner to the gallery and then eight blocks back here."

"So the diner's straight along here?"

"No sir, it's not straight along one road for any of these trips; I just meant so many edges of blocks. The city's on a square grid, and all these places are on corners, but, fact is, none of them is on the same street or avenue as any other one."

How many blocks between the library and the museum, museum and gallery, and gallery and library?

21

EVEN OR ODD

Edmund Marshall

My daughter and I once played a game based on the number 13 and the following rule:

Think of a positive whole number greater than 1. If it is even, halve it. If it is odd, multiply it by 13 and add 1. Either of these operations is to be regarded as one step. Apply another step to the outcome of the first step, and then further steps successively.

For our game, we chose different starting numbers that were odd, and the winner was the person who by the fewer number of steps reached the number 1. My daughter won because she started with the number that leads to 1 in the fewest number of steps.

What was my daughter's starting number?

RATIO

Andrew Skidmore

Liam has split a standard pack of 52 cards into three piles; black cards predominate only in the second pile. In the first pile the ratio of red to black cards is 3 to 1. He transfers a black card from this pile to the second pile; the ratio of black to red cards in the second pile is now 2 to 1. He transfers a red card from the first pile to the third pile; the ratio of red to black cards in this pile is now a whole number to one.

Liam told me how many cards (a prime number) were initially in one of the piles; if I told you which pile, you should be able to solve this teaser.

How many cards were initially in the third pile?

23

FAMILY STOCK

Howard Williams

I have been transferring shares in the family business to my grandchildren, which I've done as part of their birthday presents. On their first birthday I transferred one share, on their second birthday three shares, on their third birthday five shares etc. I have now four grandchildren and at the most recent birthday they were all of different ages.

From my spreadsheet I noticed that the numbers of shares most recently transferred to each grandchild were all exact percentages of the total number of shares transferred to all of them over their lifetimes.

In increasing order, what are the ages of my grandchildren?

24

PIE BY FIVE

Colin Vout

We had a delicious pie, rectangular and measuring 20 centimetres along the top and 13 centimetres in the other direction. We divided it into five pieces of equal area, using five straight cuts radiating from one internal point. This internal point was rather off-centre, in the top left-hand quarter, although the distances from the left and top sides were both a whole number of centimetres. The points where the cuts met the edges were also whole numbers of centimetres along the edges; one edge had two cuts meeting it, and the other three edges had one each.

How far was the internal point from the left and top sides, and how far along the four sides (starting at the top) did the cuts reach the edges (measured clockwise from the relevant corners)?

25

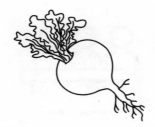

TURNIP PRIZE

Victor Bryant

The Turnip Prize is awarded to the best piece of work by an artist under fifty. This year's winning entry consisted of a mobile made up of many different plain white rectangular or square tiles hanging from the ceiling. The sides of the tiles were all whole numbers of centimetres up to and including the artist's age, and there was precisely one tile of each such possible size (where, for example, a 3-by-2 rectangle would be the same as a 2-by-3 rectangle). Last week one of the tiles fell and smashed and then yesterday another tile fell and smashed. However, the average area of the hanging tiles remained the same throughout.

How old is the artist?

26

MEMBERS OF THE JURY

Danny Roth

A jury had twelve members, all with different ages, except that two were twins with a common age over 40. The average age was a prime number. A counsel objected to one of the members, and he was replaced by another, with the result that the average age was reduced to another prime number. Between the two juries, there were twelve different ages (at least 20 but not more than 65), and they formed a progression with a common difference (e.g. 1, 4, 7, 10, 13, etc. or 6, 13, 20, 27, 34, etc.). None of the individuals had a perfect square age, and the replacement jury still included both twins.

How old were the twins?

27

LEANING TOWER OF PESOS

Howard Williams

I have six old coins worth an odd number of pesos, comprising a mixture of one- and two-peso coins. Both denominations are of the same diameter and made of the same metal, but two-peso coins are twice the thickness of one-peso coins.

After making a vertical stack of the coins I then slid each of the top five coins as far as possible to the right, to make the pile lean as much as possible in that direction, without toppling. I managed to get the rightmost edge of the top coin a distance of one and a quarter times its diameter further right than the rightmost edge of the bottom coin.

Starting from the top, what is the value of each of the six coins?

28

RECIPROCAL ARRANGEMENT

Colin Vout

Twelve men from our football squad had turned up for training, and I'd promised them a game of six-a-side at the end of the session; so while they were off on a gentle three-mile run I worked out what the two teams would be. They were wearing their squad numbers, which had one or two digits: 2, 3, 4, 5, 6, 7, 8, 9, 15, and three others. It appealed to me when I found that I could divide them into two teams of six, such that the sum of the reciprocals of the squad numbers in each team equalled one exactly.

What were the squad numbers in the team containing number 2?

29

VALUED PLAYWRIGHTS

Victor Bryant

I have given each letter of the alphabet a different whole-number value from 1 to 26. For example, P = 4, L = 8, A = 3 and Y = 24. With my numbers I can work out the value of any word by adding up the values of its letters, for example the word PLAY has a value of 39.

It turns out that the playwrights

BECKETT
FRAYN
PIRANDELLO
RATTIGAN
SHAKESPEARE
SHAW

all have the same prime value.

Also COWARD, PINERO and STOPPARD have prime values.

What are those three prime numbers?

30

FIT FOR PURPOSE

Peter Good

George and Martha bought a new toy for their son Clark. It consisted of a rectangular plastic tray with dimensions 15x16cm and eight plastic rectangles with dimensions 1x2cm, 2x3cm, 3x4cm, 4x5cm, 5x6cm, 6x7cm, 7x8cm and 8x9cm. The rectangles had to be placed inside the tray without any gaps or overlaps. Clark found every possible solution and he noticed that the number of different solutions which could not be rotated or reflected to look like any of the others was the same as his age in years.

How old was Clark?

31

FILM BINGE

Colin Vout

I'm going to have a day binge-watching films. My shortlist is: *Quaver*, starring Amerton, Dunino, Emstrey, Fairwater and Joyford; *Rathripe*, starring Amerton, Codrington, Fairwater, Glanton and Ingleby; *Statecraft*, starring Amerton, Codrington, Dunino, Hendy and Ingleby; *Transponder*, starring Codrington, Dunino, Emstrey, Hendy and Ingleby; *Underpassion*, starring Blankney, Emstrey, Fairwater, Hendy and Ingleby; *Vexatious*, starring Amerton, Blankney, Dunino, Emstrey and Joyford; *Wergild*, starring Blankney, Fairwater, Glanton, Hendy and Joyford; *X-axis*, starring Blankney, Codrington, Fairwater, Glanton and Ingleby; *Yarborough*, starring Blankney, Dunino, Glanton, Hendy and Joyford; *Zeuxis*, starring Amerton, Codrington, Emstrey, Glanton and Joyford.

I dislike Ingleby and Joyford, so I don't want either of them in more than two films; but I want to see each of the others in at least two films.

Which are the films I should watch?

32

THREE-CORNERED PROBLEM

Oliver Tailby

I have a set of equal-sized equilateral triangles, white on one face and black on the back. On the white side of each triangle I have written a whole number in each corner. Overall the numbers run from 1 to my age (which is less than thirty). If you picture any such triangle then it occurs exactly once in my set (for example, there is just one triangle containing the numbers 1, 1 and 2; but there are two triangles containing the numbers 1, 2 and 3).

The number of triangles that contain at least one even number is even. The number of triangles that contain at least one odd number is odd. The number of triangles that contain at least one multiple of four is a multiple of four.

How old am I?

33

DIAL M FOR MARRIAGE

Danny Roth

George and Martha work in a town where phone numbers have seven digits. "That's rare!" commented George. "If you look at your work number and mine, both exhibit only four digits of the possible ten (0–9 inclusive), each appearing at least once. Furthermore, the number of possible phone numbers with that property has just four single-digit prime number factors (each raised to a power where necessary) and those four numbers are the ones in our phone numbers."

"And that is not all!" added Martha. "If you add up the digits in the two phone numbers you get a perfect number in both cases. Both have their highest digits first, working their way down to the lowest."

[A perfect number equals the sum of its factors, e.g. 6=1+2+3]

What are two phone numbers?

34

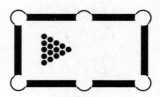

SNOOKERED

Nick MacKinnon

The playing surface of a snooker table is a twelve-foot by six-foot rectangle. A small ball is placed at P on the bottom cushion (which is six feet wide) and hit so it bounces off the left cushion, right cushion and into the top-left pocket.

Now the ball is replaced at P and hit so it bounces off the left cushion, top cushion and into the bottom right pocket, after travelling 30% further than the first shot took. The ball always comes away from the cushion at the same angle that it hits the cushion.

How far did the ball travel on the second shot?

35

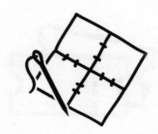

TIMELY OVERTHROWS

Howard Williams

Without changing their size, Judith sews together one-foot squares of different colours that her mother has knitted, to make rectangular throws. These are usually all of the same dimensions using fewer than a hundred squares. She has observed that it takes her mother 20 per cent longer to knit each square than it takes her to sew two single squares together.

As a one-off she has completed a square throw whose sides have the same number of squares as the longer side of her usual rectangular throws. The average time it took per square foot, both knitting and sewing, to complete the square throw was 2 per cent longer than that of the rectangular throws.

What are the dimensions in feet of the rectangular throws?

36

PRIME CUTS FOR DINNER

Peter Good

Tickets to the club dinner were sequentially numbered 1, 2, ..., etc. and every ticket was sold. The number of guests for dinner was the highest common factor of three different two-figure numbers and the lowest common multiple of three different two-figure numbers. There were several dinner tables, each with the same number of seats, couples being seated with friends. The guests on each table added their ticket numbers together and obtained one of two prime numbers, both less than 150, but if I told you the larger prime number you would not be able to determine the other.

What was the larger of the two prime numbers?

37

BEE LINES

Nick MacKinnon

Three small bees are trapped inside three empty cuboidal boxes of different sizes, none of whose faces are squares. The lengths of the edges of each box in centimetres are whole numbers, and the volume of each box is no more than a litre. Starting at a corner, each bee moves only in straight lines, from corner to corner, until it has moved along every edge of its box. The only points a bee visits more than once are corners of its box, and the total distance moved by each bee is a whole number of centimetres. Given the above, the sum of these three distances is as small as it could be.

What is the sum of the distances that the bees moved?

38

SHUFFLING SERIES SCHEDULES

Colin Vout

A TV company planned a set of programmes to fill a weekly slot (one programme per week for many weeks) with six consecutive series of three different types (Arts, Biography and Comedy). None of the series was followed by another of the same type (e.g. there could be an Arts series for three weeks then a Comedy series for four weeks and so on). Then it decided to change the order of the series within the same overall slot, but to minimise disruption it would not alter the gaps between series of the same type. It did this by scheduling each of the three Arts series 6 weeks earlier than first planned, each of the two Biography series 20 weeks later than first planned, and the Comedy series 21 weeks earlier than first planned.

How many programmes are in each of the six series after the change, in order?

39

DIGITAL DAISY-CHAINS

Victor Bryant

The number 798 is a 'digital daisy-chain'; i.e. if you spell out each of its digits as a word, then the last letter of each digit is the first letter of the next. Furthermore, the number 182 is a 'looped' digital daisy-chain because, in addition, the last letter of its last digit is the first letter of its first digit.

I have written down a large looped digital daisy-chain (with fewer than a thousand digits!). The total of its digits is itself a digital daisy-chain.

What is that total?

40

HALL OF RESIDENCE

Angela Newing

Oak Hall at Woodville University has groups of five study bedrooms per flat and they share a kitchen/diner. In one flat live language students Andy, Bill, Chris, Dave and Ed. Bill, whose home town is Dunstable is reading French. The person in room 5 comes from Colchester and Dave comes from Brighton. The chap reading German has the room with a number one greater than the man from Gloucester. Chris occupies room 3, and Ed is reading Italian. The man in room 2 is reading Spanish, and the man reading English has a room whose number is two different from the student from Reigate.

What is Andy's subject and where is his home?

41

ONE OF A KIND

Andrew Skidmore

The raffle tickets at the Mathematical Society Dinner were numbered from 1 to 1000. There were four winning tickets and together they used each of the digits from 0 to 9 once only. The winning numbers could be classified uniquely as one square, one cube, one prime and one triangular number. For example, 36 is a triangular number as 1+2+3+4+5+6+7+8 = 36, but it cannot be a winner as 36 is also a square. The tickets were all sold in strips of five, and two of the winning numbers were from consecutive strips. The smallest-numbered winning ticket was not a cube.

List the four winning numbers in ascending order.

42

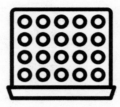

CONNECT FOUR

Howard Williams

I have four different two-digit numbers, each having at least one digit which is a three. When I multiply any three of these numbers together I get a product that, with the inclusion of a leading zero, is one or more repetitions of the repetend of the reciprocal of the fourth two-digit number. A repetend is the repeating or recurring decimal of a number. For example 1 divided by 27 is 0.037037......, giving a repetend of 037; in that case, the product would be 37 or 37037 or 37037037 etc.

What, in ascending order, are the four two-digit numbers?

43

IN THE SWIM

Nick MacKinnon

Expert mathematicians Al Gritham, Ian Tadger, Carrie Over and Tessa Mole-Poynte have a swimming race for a whole number of lengths. At some point during each length (so not at the ends) they calculate that they have completed a fraction of the total distance, and at the finish they compare their fractions (all in the lowest terms). Al says, "My fractions have prime numerators, the set of numerators and denominators has no repeats, their odd common denominator has two digits, and the sum of my fractions is a whole number." Tessa says, "Everybody's fractions have all the properties of Al's, but one of mine is closer to an end of the pool than anybody else's."

What were Tessa's fractions?

44

VILLAGE SIGNPOSTS

Colin Vout

Inside a shed I saw where the council was preparing signposts for seven neighbouring villages. Between any two villages there is at most one direct road, and the roads don't meet except at village centres, where the signposts are to stand. The arms were all affixed, and labelled except for one name to be completed on each. The names in clockwise order on each were as follows:

Barton, Aston, ?;   Barton, Grafton, ?;   Carlton, Forton, Eaton, ?;
Grafton, Aston, ?;   Dalton, Grafton, Eaton, ?;
Barton, Forton, Carlton, ?;   Dalton, Aston, ?

Starting at Dalton, I walked along roads, taking the road furthest rightwards at each village and returning to Dalton. I chose the first road so that I visited as many other villages as possible with this method.

In order, what were the other villages I visited?

45

FACE VALUE

Susan Bricket

Plato: I have written a different whole number (chosen from 1 to 9 inclusive) on each of the faces of one of my regular solids and have labelled each vertex with the product of the numbers on its adjacent faces. If I tell you the sum of those eight vertex labels, you can't deduce my numbers, but if I rearrange the numbers on the faces and tell you the new sum, then you can deduce the numbers.

Eudoxus: Tell me the new sum then.

Plato: No, but I'll tell you it's a 10th power.

Eudoxus: Aha! I know your numbers now.

Plato: Yes, that's right. But if I now told you the original sum, you couldn't work out which numbers were originally opposite each other.

What was the original sum?

LUCKY PROGRESSION

Victor Bryant

I wrote down a number that happened to be a multiple of my lucky number. Then I multiplied the written number by my lucky number to give a second number, which I also wrote down. Then I multiplied the second number by my lucky number again to give a third, which I also wrote down. Overall, the three numbers written down used each of the digits 0 to 9 exactly once.

What were the three numbers?

47

FOUR-SIDED DICE GAME

Howard Williams

I have two identical four-sided dice (regular tetrahedra). Each die has the numbers 1 to 4 on the faces. Nia and Rhys play a game in which each of them takes turns throwing the two dice, scoring the sum of the two hidden numbers. After each has thrown both dice three times, if only one of them scores an agreed total or over, then he or she is the winner. Otherwise the game is drawn.

After Nia has thrown once, and Rhys twice, they both have a chance of winning. If Rhys had scored less on his first throws and Nia had scored double her first throw total, then Nia would have had a 35 times greater chance of winning.

What are Nia's and Rhys' scores and the agreed winning total?

48

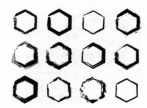

HEXAGONIA

Nick MacKinnon

Hexagonia is a hexagonal republic and is divided into 24 numbered counties, as shown. The counties are to be reorganised into four departments, each composed of six edge-connected counties. No two departments will have the same shape or be reflections of each other, and the president's residence in Department A will be built on an axis of symmetry of the department. Every sum of the county numbers in a department will be, in the prime minister's honour, a prime number, and her mansion will be built in the middle of a county in Department B, on an axis of symmetry of the department, and as far as possible from the president's residence.

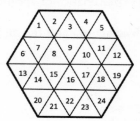

In what county will the Prime Minister's mansion be built?

49

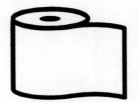

BOG STANDARD DEVIATION

Stephen Hogg

My four toilet rolls had zigzag-cut remnants (not unlike that shown). Curiously, each polygonal remnant had a different two-figure prime number of sides, each with a digit sum itself prime.

Calculating the average number of sides for the four remnants and the average of their squares, I found that the average of the squares minus the square of the average had a value whose square root (the 'standard deviation') was a whole number.

I also noticed that a regular convex polygon with a number of sides equal to the average number of sides would have an odd whole-number internal angle (in degrees).

Give the 'standard deviation'.

50

AN OLD SQUARE

Colin Vout

Archaeologists have excavated an ancient area that has the shape of a perfect square. Fragmentary documents give some information about its measurements: the number of major units along a side is less than the remaining number of minor units; there is a whole number of minor units in a major unit (no more than 10); and the area is a certain number of square major units plus a remainder of 15 square minor units.

Expressed just in terms of minor units, how many are there along a side?

MAIN LINE

Andrew Skidmore

Anton and Boris live next to a railway line. One morning a goods train passed Anton's house travelling south just as a slower train passed Boris's house travelling north. The goods train passed Boris's house at the same time as a passenger train, heading north at a speed that was half as fast again as the goods train. Similarly, as the slower train passed Anton's house it passed a passenger train; this was heading south at a speed that was three times as great as that of the slower train.

The passenger trains then passed each other at a point 25km from Anton's house before simultaneously passing the two houses.

All four trains travelled along the same route and kept to their own constant speeds.

How far apart do Anton and Boris live?

BIRTHDAY MONEY

Howard Williams

My two daughters were born on the same day but seven years apart. Every birthday, with only one year's exception, I have given them both five pounds for each year of their age. They are now grown-up, but I have continued to do this on their birthdays, except for the one year when I couldn't afford to give either of them anything.

Averaged out over all of their birthdays, including the one for which they received nothing, my elder daughter has now received 21 per cent more per birthday than her younger sister.

How much in total have I given to my daughters as birthday presents?

53

A CHRISTMAS CAROL

Nick MacKinnon

A Christmas Carol was published on 19/12/1843, when Bob Cratchit was in his forties, almost seven years after Jacob Marley's death on Christmas Eve. On Marley's last Christmas Day, a working day for them all as always, Scrooge said to Cratchit and Marley, "We three will work the same dates next year as this year, except that I'll cover Cratchit's birthday so he can have the day off, since Marley never works on his." On Boxing Day, Scrooge decided to allocate extra work dates for next year to the one whose number of work dates was going to be below the average of the three of them, bringing him up to exactly that average. Daily, up to and including New Year's Eve, Scrooge repeated this levelling-up for that person to the new average, never needing fractions of a day.

What was Bob Cratchit's full date of birth?

54

SHOUT SNAP!

Victor Bryant

My grandson and I play a simple card game. We have a set of fifteen cards on each of which is one of the words ACE, TWO, THREE, FOUR, FIVE, SIX, SEVEN, EIGHT, NINE, TEN, JACK, QUEEN, KING, SHOUT and SNAP. We shuffle the cards and then display them one at a time. Whenever two consecutive cards have a letter of the alphabet occurring on both we race to shout "Snap!".

In a recent game there were no snaps. I counted the numbers of cards between the 'picture-cards' (J/Q/K) and there was the same number between the first and second picture-cards occurring as between the second and third. Also, the odd-numbered cards (3 to 9) appeared in increasing order during the game.

In order, what were the first six cards?

55

THERE'S ALWAYS NEXT YEAR

Danny Roth

George and Martha annually examine around 500 candidates (give or take 5%). It is board policy to pass about a third of the entrants but the precise recent percentage pass rates were as follows:

Year	2018	2019	2020	2021	2022
% pass rate	30	32	35	36	X

The number of entrants in each year up to 2021 was different as was the number of successful candidates. George told Martha of the number of entries for 2022 (different again) and Martha calculated that, if X were to be once again a whole number (also different again but within the above range), the total number of successful candidates over the five-year period would be a perfect square.

How many entries were there for 2022 and how many successful candidates did Martha calculate for 2022?

56

DIDDUMS!

Stephen Hogg

'Diddums' was Didi's dada's name for numbers where the number of positive DIvisors (including 1 and the number), the number of Digits, and the Digit sUM are equal.

Didi enjoyed mental arithmetic, doing it speedily and accurately. For numbers of digits from 1 to 9 she listed in ascending order all possible numbers whose digit sum equalled the number of digits. Working through her list in order, she quickly found the first two diddums (1 and 11) and the next three diddums. After many further calculations, Didi confirmed that the sixth diddum (which is even) was listed.

"Now I'm bored," she moaned. "Diddums!" said Didi's dada.

What is the sixth diddum's digit sum?

57

THAT STRIKES A CHORD

Colin Vout

Skaredahora used a scale with seven notes, labelled J to P, for an idiosyncratic composition. It had a sequence of chords, each comprising exactly three different notes (the order being immaterial). The sequence started and ended with JNP. Each change from one chord to the next involved two notes increasing by one step in the repeating scale, and the other decreasing by two steps (e.g. JNP changing to KLJ).

Every chord (eventually) reachable from JNP was used at least once; every allowable change from one of these chords to another was used exactly once. The number of chord changes between the first pair of JNPs was more than that between the last such pair, but, given that, was the smallest it could be.

With notes in alphabetical order, what was the seventh chord in the sequence?

58

CRAZY GOLF

Angela Newing

Ian was trying to entertain John and Ken, his 2 young nephews, in the local park, which had a 9-hole crazy golf course. He decided to play a round with each of them separately, and gave them money for each hole he lost on a rising scale. He would pay £1 if he lost the first hole (or the first hole after winning one), then £2 if he lost the next consecutive hole, and £3 for the third, and so on.

In the event, Ian won only 5 holes in total between the two rounds, including the first hole against John and the last hole against Ken. There were no ties. At the reckoning after both rounds, both boys received equal amounts of money.

How much did it cost Uncle Ian?

59

WONKY DICE

Michael Fletcher

I have two octahedral (eight-faced) dice. There are positive whole numbers (not necessarily all different) on each of their faces. I throw the two dice and add the two numbers that are showing on the top faces. The probability of a total of 2 is 1/64, the probability of a total of 3 is 2/64 and so on. The probabilities of the totals are the same as for a pair of standard octahedral dice (each numbered 1 to 8). Interestingly, however, the highest number on the first die is 11.

What are the eight numbers (in ascending order) on the second die?

60

RECURRING THEME

Andrew Skidmore

Liam is learning about fractions and decimals. He has been shown that some fractions give rise to decimals with a recurring pattern, e.g. 4/11 = 0.3636... . Each pupil in his class has been given four numbers. The largest (two-figure) number is to be used as the denominator and the others as numerators, to create three fractions that are to be expressed as decimals. Liam found that each decimal was of the form 0.abcabc..., with the same digits but in different orders.

Liam's largest number contained the digit 7 and if I told you how many of his four numbers were divisible by ten you should be able to work out the numbers.

What, in increasing order, are the four numbers?

61

CRATE EXPECTATIONS

Nick MacKinnon

Pip and Estella know each other to be honest and clever. At the reading of Miss Havisham's will they see a large crate on which are placed two envelopes addressed V+S+E and V, and a lipstick. The solicitor gives Estella the V+S+E envelope and Pip the V envelope and says, "Inside this crate Miss Havisham has left a cuboidal gold bar whose dimensions in mm are different whole numbers greater than 1, and whose volume, surface area and total edge length are V mm³, S mm² and E mm respectively. Inside your envelope, which you should now open, is the secret value of your address. Every fifteen minutes a bell will ring, and if at that point you know the other's value, write both values on the crate with lipstick and the gold is yours." At the first bell Pip and Estella sat still, but when the bell rang again Estella lipsticked 3177 and Pip's value on the crate.

What was Pip's value?

62

LAP PROGRESSION

Mark Valentine

In a charity fundraising drive, two friends completed laps of their local track. Alan, a cyclist, raised £1 for his first lap, £2 for the second, £3 for the third and so on, with each extra lap earning £1 more than the prior. Bob, who walks, raised £1 for his first lap, £2 for the second, £4 for the third and so on, with each extra lap earning double the prior. Starting together, they travelled at constant speeds, and finished their last lap simultaneously.

After they had added up their totals, they were surprised to find they each raised an identical four-figure sum.

Including the beginning and end of their drive, how many times did Alan and Bob cross the lap start-finish line together?

63

JOKERS

Victor Bryant

A group of us wanted to play a card game. We had a standard 52-card pack but we needed to deal out a whole pack with each person getting the same number of cards, so I added just enough 'jokers' to make this possible. Then I randomly shuffled the enlarged pack (in this shuffled pack there was more than a 50:50 chance that there would be at least two jokers together). Then I dealt the cards out, each of us getting the same number (more than three). There was more than a 50:50 chance that my cards would not include a joker.

How many of us were there?

64

EMPOWERED

Howard Williams

I have a rectangular garden whose sides are whole numbers of feet. Away from the edge, an exposed strip of ground, again a whole number of feet in width, runs straight across (not diagonally) from a shorter side of the garden to the other shorter side. I need to run an electrical cable along the ground, between two opposite corners of the garden. Where the cable crosses the exposed area, it has to be threaded through expensive linear ducting to avoid damage. Because of costs, whilst seeking to minimise the length of cable, my overriding concern is to minimise the length of ducting used.

The straight-line distance between the two corners to be connected is 123 feet, but in minimising costs, the length of cable needed is a whole number of feet longer than this.

What is the length of cable needed?

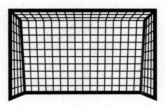

PRIME FOOTBALL

Edmund Marshall

There are more than 13 clubs in the Prime Football League, and each club plays each other twice during the football season, gaining three points for winning a match, and one point for a draw.

At the end of last season, it was observed that the final numbers of points gained by the clubs were all different double-digit prime numbers.

During the season our local club, Wessex Wanderers, won just three times as many matches as they lost. If you knew how many clubs were in the league, you could work out the final points total of Wessex Wanderers.

How many clubs are in the league, and what was the final points total of Wessex Wanderers?

66

ROMAN PRIMES

Peter Good

Romulus played a game using many identical six-sided dice. Each die face shows a different single-character Roman numeral. He rolled two dice and was able to form a prime number in Roman numerals by arranging the two characters. Romulus rolled more dice (one at a time) and, after each roll, formed a prime number with one more character, rearranging the sequence as needed, until he could no longer form a prime number less than 100. He was using standard notation where a character before another that is 5 or 10 times larger is subtracted from the larger one, e.g. IX=9.

After playing many games he realised that there were exactly three primes less than 100 that could not occur in any of the sequences.

In decimal notation and in ascending order, what were those prime numbers?

67

UNDERGROUND UMBRA CONUNDRUM

Stephen Hogg

An underground car park has rectangular walls, each covered with 300 abutting square tiles arranged in three repeats of red, orange, yellow, green and blue columns twenty tiles high. A sharp tapering shadow from a duct played across one wall from top to bottom (with straight lines similar to that shown).

Each shadow edge crossed every colour and, curiously, hit just two points where four tile corners met. The upper and lower pairs of these were on levels the same distance up and down from the mid-level line. I also noticed that no column was completely free of shadow, but just one column was completely in shadow.

Give the shadow's area relative to the wall area as a fraction in simplest form.

ROOM 101

Nick MacKinnon

In the Ministry of Love, O'Brien told Winston Smith to calculate five-digit perfect squares, none sharing a prime factor with 1984. "And Winston, no two of them may differ by a multiple of 101 or you know what will happen. Now find as many as you can." When Winston said he had found more than fifty, O'Brien replied, "Well done, Winston. Although there were a quadrillion solutions, I know some of your squares." "Does the party control my mind?" "We do, Winston. Did you choose 10201, for example?" "Of course not! It's the worst square in the world!" "How many fingers am I holding up, Winston?" "Three? Four? I don't know!" "That's how many of your squares we know."

What four squares did O'Brien know?

69

MY GRANDFATHER'S COINS

Bernardo Recamán

When my grandfather died, he left his fine collection of coins, not more than 2500 of them, to his children, a different number to each of them, and in decreasing amounts according to their ages.

To the eldest of his children, he left one fifth of the coins, while the youngest inherited just one eleventh of them. Gaby, my mother, and third oldest of the children, received one tenth of the collection. Each of the other children received a prime number of coins.

How many coins did my aunt Isabel, second oldest in the family, inherit?

70

HOLE NUMBERS

Colin Vout

In theoretical golf, you have a set of "clubs" each of which hits a ball an exact number of yards forwards or backwards. You own the following set: 3, 8, 17, 19 and 35. For example, if a hole is 31 yards long, you can reach it in three strokes, with two forward hits of the 17 and a backward hit of the 3. In the next competition, you are only allowed to use three clubs, and the course consists of three holes whose lengths are 101, 151 & 197 yards. In order to get round the course in the fewest possible strokes, you must make a wise choice of clubs.

Which three clubs (in ascending order) should you choose, and what will the individual hole scores be (in order)?

71

MANY A SLIP

Victor Bryant

I have written down three 3-figure numbers in decreasing order and added them up to give their 4-figure sum, which is a perfect square: the digit 0 occurred nowhere in my sum.

Now I have attempted to replace digits consistently by letters and I have written the sum as

CUP + AND + LIP = SLIP

However, there's "many a slip twixt cup and lip" and unfortunately one of those thirteen letters is incorrect. If you knew which letter was incorrect then you should be able to work out the three 3-figure numbers.

What are they?

72

DON'T MISS A SECOND

Howard Williams

I have an analogue wall clock with a second hand and also a separate 24-hour hh:mm:ss digital clock. The wall clock loses a whole number of seconds over a two-digit period of seconds. The digital clock gains at a rate 2½% greater than the wall clock loses. After resetting both clocks to the correct time, I noticed that they both displayed the same but wrong time later in the same week, at a time of day one hour earlier than that when they were reset.

I can reset one of the clocks at an exact hour so that it will show the correct time when the televised rugby kicks off at 19:15:00 on the 31st.

What is the latest time (hour and date) when I can do this?

73

PIN

Andrew Skidmore

Callum has opened a new current account and has been given a telephone PIN that is composed of non-zero digits (fewer than six). He has written down the five possible rearrangements of his PIN. None of these five numbers are prime; they can all be expressed as the product of a certain number of different primes.

The PIN itself is not prime; it can also be expressed as the product of different primes but the number of primes is different in this case. The sum of the digits of his PIN is a square.

What is the PIN?

74

THE PLUMBER'S BUCKETS

Peter Good

A plumber was trying to empty a tank containing 100 litres of water using three buckets, each marked with a different whole number of litres capacity between 10 and 20 litres. He calculated that he could exactly empty the tank, but only by using all three buckets. He then emptied the tank as calculated, filling each bucket a different number of times, but then found that the tank still contained 6 litres of water, because the smallest bucket had a dent that reduced its capacity by 3 litres.

What were the marked capacities of the three buckets?

75

COLOURFUL CHARACTERS

Danny Roth

George and Martha have recently taken a great-grandchild to a toddler's birthday party. The youngsters like to traipse around over a pen with a large number of brightly coloured plastic balls. Actually there were 200 in total, some of red, yellow, blue and green. There were at least 30 but fewer than 70 of each colour, with the following properties:

Red	perfect square
Yellow	prime number
Blue	palindromic number
Green	divisible by three single-digit prime numbers

George told Martha the above information and the number of red balls. Martha was then able to work out the numbers of each of the others.

How many of each colour were there?

76

GERMOMETRIC MEAN

Stephen Hogg

On Whit Monday, Zak began self-isolating upstairs. At lunchtime Kaz shouted up, "What's a Geometric Mean?" "It's the Nth root of the product of N values," Zak replied.

On TV, Teaseside hospital's 'geovid' admissions for the seven days prior were listed alongside their Geometric Mean. Kaz stated that chronologically the numbers comprised a decreasing set of two-figure values, Friday's value equalling the Geometric Mean. She added that, curiously, there was a value double the Geometric Mean, but not triple, whereas the Geometric Mean was triple a data value, but not double a data value. She then told Zak just the Geometric Mean.

Zak worked out the unique data set.

Give the seven numbers in chronological order.

77

POLL POSITIONS

Nick MacKinnon

In an election for golf-club president, voters ranked all four candidates, with no voters agreeing on the rankings. Three election methods were considered.

Under First-past-the-post, since the first-preferences order was A, B, C, D, the president would have been A.

Under Alternative Vote, since A had no majority of first preferences, D was eliminated, with his 2nd and 3rd preferences becoming 1st or 2nd preferences for others. There was still no majority of 1st preferences, and B was eliminated, with his 2nd preferences becoming 1st preferences for others. C now had a majority of 1st preferences, and would have been president.

Under a Borda points system, candidates were given 4, 3, 2, or 1 points for each 1st, 2nd, 3rd or 4th preference respectively. D and C were equal on points, followed by B then A.

How many Borda points did each candidate receive?

78

STOP ME IF YOU'VE HEARD THIS ONE

Colin Vout

A square, a triangle and a circle went into a bar. The barman said, "Are you numbers over 18?" They replied, "Yes, but we're under a million." The square boasted, "I'm interesting, because I'm the square of a certain integer." The triangle said, "I'm more interesting; I'm a triangular number, the sum of all the integers up to that same integer." The circle said, "I'm most interesting; I'm the sum of you other two." "Well, are you actually a circular number?" "Certainly, in base 1501, because there my square ends in my number exactly. Now, shall we get the drinks in?" The square considered a while, and said, "All right, then. You(')r(e) round!"

In base 10, what is the circular number?

79

PRODUCT DATES

Edmund Marshall

If a date during the current millennium is day D in month M during year (2000+N), it is said to be a product date if the product of D and M equals N (for example 11 February 2022). My daughter and I have been investigating the numbers of days from one product date to the next product date. I was able to establish the longest such interval L, while my daughter worked out the shortest such interval S. We were surprised to find that L is a whole number multiple of S.

What is that multiple?

80

HIDDEN POWERS

Victor Bryant

My grandson is studying "History since the Battle of Hastings". I made him a game, which consisted of a row of nine cards, each with a different non-zero digit on it. Throw a standard die, note the number of spots displayed, count that number of places along the row and pause there. Throw the die again, move the corresponding number of places further along and pause again. Repeat this until you come off the end of the row, noting the digit or digits you paused on and put these together in the same order, to produce a number.

Keeping the cards in the same order I asked my grandson to try to produce a square or cube or higher power. He eventually discovered that the lowest possible such number was equal to the number of one of the years that he had been studying.

What is the order of the nine digits along the row?

81

BUS STOP BLUES

Susan Bricket

While waiting for buses, I often look out for interesting number plates on passing cars. From 2001 the UK registration plate format has been 2 letters + a 2-digit number + 3 more letters, the digits being last two of the year of registration with 50 added after six months (for example in 2011, the possible numbers were 11 and 61). I spotted one recently with its five letters in alphabetical order, all different and with no vowels. Looking more closely, I saw that if their numerical positions in the alphabet (A = 1, B = 2 etc.) were substituted for the 5 letters, their sum plus 1 was the 2-digit number and the sum of their reciprocals was equal to 1.

Send the 7-character registration.

82

TOP MARKS

Howard Williams

A teacher is preparing her end-of-term class test. After the test she will arrive at each pupil's score by giving a fixed number of marks for a correct answer, no marks if a question is not attempted, and deducting a mark for each incorrect answer. The computer program she uses to prepare parents' reports can only accept tests with the number of possible test scores (including negative scores) equal to 100.

She has worked out all possible combinations of the number of questions asked and marks awarded for a correct answer that satisfy this requirement, and has chosen the one that allows the highest possible score for a pupil.

What is that highest possible score?

83

BANK ROBBERY

Angela Newing

Five witnesses were interviewed following a robbery at the bank in the High Street. Each was asked to give a description of the robber and his actions. The details given were: height, hair colour, eye colour, weapon carried, escape method.

Witness 1: short, fair, brown, cricket bat, motor bike
Witness 2: tall, fair, grey, gun, car
Witness 3: tall, dark, brown, crowbar, motor bike
Witness 4: short, ginger, blue, knife, car
Witness 5: tall, dark, blue, stick, push bike

When the police caught up with the perpetrator, they found that each of the five witnesses had been correct in exactly two of these characteristics.

What was the robber carrying, and how did he get away?

84

A SIX-PIPE PROBLEM

Nick MacKinnon

A factory makes six types
of cylindrical pipe, A to F
in decreasing size, whose
diameters in centimetres are
whole numbers, with type A
50% wider than type B. The
pipes are stacked in the yard

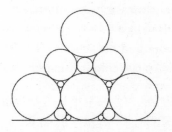

as a touching row of As with an alternating row of touching Bs
and Cs in the next layer, with each B touching two As. Type Ds fill
the gap between the As and the ground; Es fill the gap between
As and the Bs; and Fs fill the gap between As, Ds and the
ground. Finally another row of As is put on top of the stack,
giving a height of less than 5 metres.

What is the final height of the stack in centimetres?

85

LAWN ORDER

Colin Vout

A gardener was laying out the border of a new lawn; he had placed a set of straight lawn edging strips, of lengths 16, 8, 7, 7, 7, 5, 4, 4, 4 & 4 feet, which joined at right angles to form a simple circuit. His neighbour called over the fence, "Nice day for a bit of garden work, eh? Is that really the shape you've decided on? If you took that one joined to its two neighbours, and turned them together through 180°, you could have a different shape. Same with that one over there, or this one over here – oh, look, or that other one." The gardener wished that one of *his* neighbours would turn through 180°.

What is the area of the planned lawn, in square feet?

THE BEARINGS' TRAIT

Stephen Hogg

At Teaser Tor trig. point I found a geocaching box. The three-figure compass bearings (bearing 000=north, 090=east, etc.) from there to the church spires at Ayton, Beeton and Seaton were needed to decode the clue to the next location.

Each spire lay in a different compass quadrant (e.g. 000 to 090 is the North-East quadrant). Curiously, each of the numerals 1 to 9 occurred in these bearings and none of the bearings were prime values.

Given the above, if you chose one village at random to be told only its church spire's bearing, it might be that you could not calculate the other two bearings with certainty, but it would be more likely you could.

Give the three bearings, in ascending order.

SWEET SUCCESS

Victor Bryant

My five nieces Abby, Betty, Cathy, Dolly and Emily each had some sweets. I asked them how many they had but they refused to answer directly. Instead, in turn, each possible pair from the five stepped forward and told me the total number of sweets the two of them had. All I remember is that all ten totals were different, that Abby and Betty's total of 8 was the lowest, and that Cathy and Dolly's total of 18 was the second highest. I also remember one of the other totals between those two but I don't remember whose total it was. With that limited information I have worked out the total number of sweets.

In fact it turns out that the other total I remember was Betty and Cathy's.

In alphabetical order of their names, how many sweets did each girl have?

88

PUTTING IN A SHIFT

Bill Kinally

Next month's four-week rota for Monday to Friday dinner duties starting on Monday 1st is covered by five teachers each having the following duty allocations. Ann, Ben and Ed each have four, Celia six and Dave has two. Strangely, all the prime number dates (1 is not a prime) are covered by Ben, Celia and Dave, while the others are covered by Ann, Celia and Ed. After working a duty, nobody works on either of the following two shifts, so anyone working on a Friday will not work on the following Monday or Tuesday. Celia has no duties on Mondays while Ben and Ed have none on Wednesdays.

In date order, who is on duty from Monday to Friday of the first week?

89

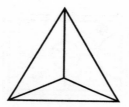

TETRAHEDRAL TOWERS

Edmund Marshall

I have a large number, fewer than 2000, of identical spherical bonbons, arranged exactly as a tetrahedral tower, having the same number of bonbons along each of the six edges of the tower, with each bonbon above the triangular base resting on just three bonbons in the tier immediately below.

I apportion all my bonbons between all my grandchildren, who have different ages in years, not less than 5, so that each grandchild can exactly arrange his or her share as a smaller tetrahedral tower, having the same number of tiers as his or her age in years.

The number of my grandchildren is the largest possible in these circumstances.

How many tiers were in my original tower, and how old in years are the eldest and youngest of my grandchildren?

90

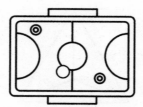

BOUNCE COUNT

Mark Valentine

At the local arcade, Claire and David played an air hockey game, consisting of a square table with small pockets at each corner, on which a very small puck can travel 1m left-right and 1m up-down between the perimeter walls. Projecting the puck from a corner, players earn a token for each bounce off a wall, until the puck drops into a pocket.

In their game, one puck travelled 1m farther overall than its left-right distance (for the other, the extra travel was 2m). Claire's three-digit number of tokens was a cube, larger than David's number which was trianglular (1+2+3+...). Picking up a spare token, they could then arrange all their tokens into a cube and a square combined.

How many tokens did they end up with?

91

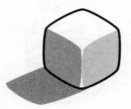

MAKING SQUARES

Andrew Skidmore

Liam has nine identical dice. Each die has the usual numbers of spots from 1 to 6 on the faces, with the numbers of spots on opposite faces adding to 7. He sits at a table and places the dice in a 3x3 square block arrangement.

As I walk round the table I see that (converting numbers of spots to digits) each vertical face forms a different three-figure square number without a repeating digit.

As Liam looks down he sees six three-digit numbers (reading left to right and top to bottom) formed by the top face of the block, three of which are squares. The total of the six numbers is less than 2000.

What is that total?

92

HEADS UP SAVINGS

Howard Williams

Little Spencer saves 5p coins in a jar, and when they reach £10, deposits them in his savings account.

He likes playing with the coins. In one game, after first turning them all heads up, he places them in a row on the table. Starting from the left, he then turns over every 2nd coin until he has reached the end of the row. He then again starts from the left, and this time turns over every 3rd coin. He repeats this for every 4th, 5th coin etc., until finally he turns over just one coin, the last in the row.

At the end of the game I could see that if Spencer had exactly 75% more coins he would have an increase of 40% in the number showing heads. However, if he had exactly 50% fewer coins, he would have a decrease of 40% in the number showing heads.

What is the value of his 5p savings?

BILATERAL SYMMETRY

Susan Bricket

My son, at a loose end after A-Levels, asked me for a mental challenge. As we'd been discussing palindromes, I suggested he try to find an arrangement of the digits 1 to 9 with the multiplication symbol 'x' to give a palindrome as the answer, and he came up with 29678 x 1453 = 43122134. I said: "Now try to find the smallest such palindromic product starting in 4, where the last digit of the smallest number is still 3". He found that smallest product, and, interestingly, if he added the two smaller numbers instead of multiplying them, then added 100, he also got a palindrome.

What was the smallest product?

94

PRIMARY ROAD NETWORK

John Owen

I was recently studying a large map that showed all the towns and major roads in a country. Every town had at least one road leading to it and every road led from one town to another. The roads only met at towns and all joined together to make a network with lots of blank areas in between, which I happily coloured in, needing just four different colours.

I counted up the number of areas (excluding the area around the outside of the network) and the number of towns, and discovered that both numbers were prime. Also, when I took these two numbers with the number of roads, the three numbers together used every digit from 0 to 9 precisely once.

In increasing order, what were the three numbers?

95

STRINGING ALONG

Peter Good

An artist hammered thin nails from a pack of 40 into a board to form the perimeter of a rectangle with a 1cm gap between adjacent nails. He created a work of art by stringing a long piece of wire from one nail to another, such that no section of wire was parallel to an edge of the rectangle. The wire started and ended at two different nails, no nail was used more than once and the length of the wire was a whole number of cm. No longer wire was possible that satisfied the above conditions.

What were the dimensions of the rectangle and the length of the wire chain (in cm)?

96

RHYTHMIC GYMNASTICS

Colin Vout

Skaredahora used three rhythmic patterns of quaver beats in a short, purely percussive, composition. His 'Dotykam' rhythm has accents on beats 1, 4, 6, 7, 8, 10 & 11; 'Kluc' has accents on beats 1, 8 and one particular beat in between; and 'Omacka' has accents on beats 1, 2, 5, 6 & 10. Several percussion instruments are involved, each playing one of the three rhythms, but starting at different times. Overall the patterns overlap, with every beat in the composition being filled by an accent from exactly one of the instruments, and all the patterns finishing by the end of the composition.

What is the other beat of Kluc, and what is the order of appearance of the rhythmic patterns (e.g. DOOKD)?

97

FILL CELLS MENTALLY,
MY DEAR WATSON

Stephen Hogg

Moriarty's papers were alight. Holmes memorised a 3x3 grid of different 4-digit values in ascending order as illustrated, from A (lowest) to I (highest), noticing that one numeral wasn't used. Three cells, isolated from one another (no corner or

A	B	C
D	E	F
G	H	I

edge contact), held squares of squares. Three others, similarly isolated, held cubes of non-squares. The other three held squares of non-squares. Holmes told Watson these features, specifying only the lowest value. "Not square," remarked Watson.

"True, and many grids match these facts. However, if I told you the positions of the squares of squares in the grid, you could deduce a majority of the other eight values (apart from the lowest one)" replied Holmes.

In ascending order, which values did Watson now know certainly?

98

COMMON NAMES

Victor Bryant

Eight friends met at a party; their ages in whole numbers of years were all different. They were Alan, Cary, James, Lucy, Nick, Ricky, Steve and Victor, with Lucy being the youngest. For each of them the square of their age was a three-figure number consisting of three different digits. Furthermore, for any two of them, the squares of their ages had at least one digit in common precisely when their names had at least one letter in common.

In alphabetical order of their names, what are the eight ages?

PRIMUS INTER IMPARES

Edmund Marshall

Just nine Prime Ministers held office in the Kingdom of Primea during the twentieth century. No two Prime Ministers held office at the same time, none had more than one period in office, and the gap between successive Prime Ministers' terms was never more than a month. Each held office for a period of time in which the number of whole years was a different prime number (e.g. holding office from 1910 to 1915 could cover 4 or 5 whole years) and no Prime Minister served for more than 30 years. Appropriately, they all took office in prime years, but there was no change of Prime Minister in 1973.

In which years during the twentieth century did new Prime Ministers in Primea take up office?

100

CHECKMATE

Andrew Skidmore

Our chess club is divided into sections, with each section having the same number of players. The two oldest members will soon be retiring from playing and we will change the number of sections. The number of sections will change by one and so will the number of players in each section, but all the sections will still have the same number of players. This will result in there being a grand total of 63 fewer matches per year if each member plays all the other members of their section once per year.

How many players are in each section at the moment?

SOLUTIONS

1   MOVING DIGIT

Answer: 31589712

We have ab....yzn $*$ n = nab.....yz, where n > 1

The pin number may be found by considering values of n from 2 to 9 and working from right to left to fill in the values of the digits. We require b = 0 (then a = 1) so the process for each n value can be terminated if no zero appears after five digits.
In this way we find n = 4: 102564 $*$ 4 = 410256

All other values of n lead to pin numbers with greater than ten digits.

410256 = $(2^4) (3^2) (7) (11) (37)$

The correct set of three numbers for the sort code must have values close to each other to give the smallest sum. 74 must be present rather than 37 and the three numbers must be 74, 72 and 77.
The account number cannot contain 4 (the digit which moves).

410256 x 74 = 30358944   410256 x 72 = 29538432   410256 x 77 = **31589712**

2 THE BEST GAME PLAYED IN 1966

Answer: 47

1st. shot – find prime values of 200-P >100 – for 3<=P<=97 – gives following valid options

P	97	73	61	43	37	19	7	3
200-P	103	127	139	157	163	181	193	197

2nd. shot – from these take prime Q (not=P) from 2<=Q<(200-P)/2 to give a square. For Q=2 – no squares. For odd Q, (200-P)-Q is even so only consider 64, 100 and 144 when allowed.

200-P	103	127	139	157	163	181	193	197
Q	3	-	-	13	19	37	-	53;97
200-P-Q	100			144	144	144		144;100

3rd. shot – for prime R (not=P or Q) – 100-R>50 is 98 or lower odd, only cube=64 – invalid

For 144-R>72 only cube=125 so **R=19** (and so P, Q not=19)

4th. shot – for prime S (not=P, Q, R[=19]) 125-S=(square>1)x(prime)>62.5, so 2<=S<=61

S	2	3	5	7	11	13	17	23	29
125-S	123	122	120	118	114	**112**	**108**	102	96
S	31	37	41	43	47	53	59	61	
125-S	94	88	84	82	78	72	66	64	

Valid options: **S=13, 125-S=112=16x7; S=17, 125-S=108=36x3; S=53; 125-S=72=36x2**

5th. shot – for prime T (not=P, Q, R[=19], S), (125-S)-T=(cube>1)x(prime)>(125-S)/2
For T=2, 125-S-T=110, 106 or 70 all invalid, so 125-S-T=(cube)x(prime) is odd(<125). Only cube option=27 and only prime option=3. So 125-S-T=81=27x3 – [125-S=108=81+27 and 125-S=72 both invalid] – **so 125-S=112=81+31 giving S=13 and T=31**

With **R=19, S=13** three of the 2nd. shot 144 options are eliminated resulting from P=19 or Q=13 or Q=19 – see above. So 200-P-Q=144 results from **P=3 and Q=53**

6th. shot – so hit sequence:- **200[-3]=197[-53]=144[-19]=125[-13]=112[-31]=81[-U]=** not prime, square, cube, (square>1)x(prime) or (cube>1)x(prime)

81-U>40.5, so prime U<41 (not=P[=3], Q[=53], R[19], S[=13], T[=31])

U	2	5	7	11	17	23	29	37
81-U	79	76	**74**	**70**	64	**58**	52	44
Factorize	prime	4x19	**2x37**	**2x5x7**	2^6	**2x29**	4x13	4x11

7th. shot – so for prime V (not=U, 3, 13, 19, 31) <37 or <29 if 81-U=58, find 81-U-V=prime

(81-U); V=	2	5	7	11	17	23	29
74 (U=7)	72	69	na	63	57	51	45
70 (U=11)	68	65	63	na	**53**	**47**	**41**
58 (U=23)	56	**53**	51	**47**	**41**	na	-

Valid U, V swaps for U=11 or 23. So only U=11;23, V=23;11 – **47 blocks left**

3 A QUESTION OF SCALE

Answer: 100010101110

Methodically considering possibilities will get you to the solution, and there is a point that simplifies this process.

Any scale satisfying the basic rules could be started anywhere in the repeating cycle, and/or the order could be reversed, to give another set that satisfies the rules. This means that for initial analysis you can conduct your search in an organised way that 'without loss of generality' examines just one scale from each 'equivalent' related set. Then, when you have found an answer in this way, you can consider all the scales related to it (by 'rotation and reflection') and identify which is the desired solution by applying the extra rule for Skaredahora's favouritism.

This allows you to organise an exhaustive search by considering all the ways to have 2 notes in the set having another exactly 1 above, and then seeing if it is possible to have exactly 1 with another one 5 above. Without loss of generality, this gives the following patterns; firstly the ones where the 2 pairs 1 apart have no notes in common, and then those where they do have a note in common.

- 110110?????0 needs 2 more notes, but there isn't enough room for them to be 5 apart; so one of the extra notes must be 5 from a note already there, which gives just these non-equivalent possibilities:
  - 11011010???0, but then each of the unknown positions for the last note is 5 away (up or down) from a note already present, so a 6-note scale can't be formed
  - 110110?010?0, but again the last note can't be placed legally
  - 110110??0100, which again can't be completed
- 1100110????0 already has a pair of notes 5 apart, and all the unknown positions are 5 away from an existing note, so the scale can't be formed
- 110?0110???0 has two pairs of notes 5 apart, so already the rules have been broken
- 110??0110??0 also has two pairs of notes 5 apart, so already the rules have been broken
- 1110???????0 needs 3 more notes, but of the unknown positions it is only the 2 end ones that are not 5 away (either up or down) from the existing 1s; therefore exactly one of the middle four unknown positions must be a 1, and then the 2 end ones must also be 1s; therefore there are only the following non-equivalent possibilities:
  - 111011000010, which has another pair 1 apart, which breaks the rules
  - 111010100010, which works successfully
  - 111010010010, which has a note that is 5 away from 2 existing notes, which breaks the rules

Of all the rotations and reflections of the scale found to satisfy the rules, the numerically earliest one that begins with a 1 is 100010101110, which is therefore Skaredahora's favourite scale.

4 AN ODD SELECTION

Answer: 5159 and 5951

The sum of the odd digits in each column will be 16, 18, 20, 22 or 24. Also, as each digit is missing from at most one column, the sum of each column will be different.

The sum of the four across numbers lies between 4x5111=20444 and 4x5999=23996 so that restricts the possibilities for the sums of the first two columns. Furthermore, we can rule out 20 16 and 20 24 for those first two sums because they would always give a total which, when divided by four, was 54-- or 56--. So the first two columns must have sums 18 20 or 18 22 or 18 24 or 20 18 or 20 22 or 22 16 or 22 18 (followed by) 16

Also, in order to get an odd whole-number average, the sum of the four across numbers is divisible by 4 but not by 8. So, for example, the fourth column cannot add to 18 as it would result in the sum of the across numbers ending ...*odd* 8, not divisible by 4. Similarly, the last two columns adding to 20 and 16 respectively would give a sum ... *even* 16, divisible by 8. We can soon see that the last two columns must add to one of: 18 16 or 22 16 or 16 20 or 24 20 or 18 24 or 22 24.

Combining these two sets of results we get the possible values for the four column sums listed below, grouped together by the total of all the digits in the grid:

Total of all the digits in the grid	Column sums in order	Sum of four across numbers	Average of the four
76	20 18 22 16	22036	550-
	20 22 18 16	22396	5599
	22 18 16 20	23980	5995
78	18 24 16 20 only		
80	18 24 22 16	20636	**5159**
	22 16 18 24	23804	**5951**
82	22 16 24 20 only		
84	18 22 24 20	20460	5115
	20 18 22 24	22044	5511
	20 22 18 24	22404	56--

Of course precisely the same logic can be applied to the four four-figure down numbers and so we're looking for two different odd-digited averages in the same category above. In all cases, except the bold ones, the averages use an even number of different odd digits. So the averages are 5159 and 5951:
e.g.

1	9	7	5
3	5	1	7
5	3	9	1
9	7	5	3

5 BUILDING BLOCKS

Answer: 45, 91 & 990

This is of course to do with what are known as triangular numbers, i.e. the sums of all the whole numbers up to n; the formula for that is $\frac{1}{2}n(n+1)$, with the first few values being
1, 3, 6, 10, 15.

To find possible numbers of red and blue blocks, you can consider all the values up to 100 (which corresponds to $n \leq 13$), and check which pairs of different numbers sum to another triangular number. This gives 6+15=21, 10+45=55, 15+21=36, 21+45=66, 36+55=91, and 45+91=136. Written as the nth triangular numbers, these are $3^{rd}+5^{th}=6^{th}$, $4^{th}+9^{th}=10^{th}$, $5^{th}+6^{th}=8^{th}$, $6^{th}+9^{th}=11^{th}$, $8^{th}+10^{th}=13^{th}$, and $9^{th}+13^{th}=16^{th}$.

Discovering how there could be a suitable number of yellow blocks will be laborious unless you apply a bit of simple algebra. To have two triangular numbers adding up to a third triangular number you need to find integer solutions to $\frac{1}{2}a(a+1) + \frac{1}{2}b(b+1) = \frac{1}{2}c(c+1)$. You can rewrite this equivalently as $a(a+1)=c(c+1) - b(b+1)=c^2-b^2+c-b=(c-b)(c+b+1)$. So what you need to do is find two numbers $x=c-b$ and $y=c+b+1$ that multiply to make $a(a+1)$; then c solves as $\frac{1}{2}(y+x-1)$ and b solves as $\frac{1}{2}(y-x-1)$. For this to give positive whole number results, one of x and y must be even and the other must be odd, and y must be larger than $x+1$. There will typically be a few possibilities for this.

The possible pairs of nth triangular numbers are given above, describing how many red (say the pth triangular number) and blue (say q^{th}) blocks there might be. Obviously, what you need to do now is go through all the pairs and see if you can find a number of yellow blocks (say the rth triangular number) that would solve the problem for one of the pairs. To do this, for each pair you need to put $a = p$ in the formula above and find the set of potential b values, and then put $a = q$ in the formula and find another set of potential b values, and see if there is one potential b that is in both sets; this would give a value for r that works.

For speed you can write out the working as $a \rightarrow a(a+1) \rightarrow x,y \rightarrow b,c$. This gives:
3 → 12 → 1,12 → 5,6
5 → 30 → 1,30 or 2,15 or 3,10 → 14,15 or 6,8 or 3,6
4 → 20 → 1,20 → 9,10
9 → 90 → 1,90 or 2,45 or 3,30 or 5,18 or 6,15 → 44,45 or 21,23 or 13,16 or 6,11 or 4,10
6 → 42 → 1,42 or 2,21 or 3,14 → 20,21 or 9,11 or 5,8
8 → 72 → 1,72 or 3,24 → 35,36 or 10,13
10 → 110 → 1,110 or 2,55 or 5,22 → 54,55 or 26,28 or 8,13
13 → 182 → 1,182 or 2,91 or 7,26 → 90,91 or 44,46 or 9,16

When a=3 and then a=5 there are no b numbers in common. Similarly for 4 & 9, and 5 & 6, and 6 & 9, and 8 & 10. But for 9 & 13 there is b=44 in common. So the solution to the question is the 9th, 13th and 44th triangular numbers, i.e. 45, 91 and 990.

6 LET TEL PLAY BEDMAS HOLD 'EM!

Answer: +1274, +1289

Interspersing four 'operator' cards (none together) among the 'ace' to '9' needs five values, making four gaps. A value>1000 means one 4-fig. or one 5-fig. number – options as follows:-

12345	6	7	8	9	or	1	23456	7	8	9	or
1	2	34567	8	9	or	1	2	3	45678	9	or
1	2	3	4	56789	or	All invalid – more odd than even numbers					

1234	**5÷**	**6x[x]**	**78[+]**	**9**	*	or	1234	5	6	7	89	invalid	or
1234	**56÷[x]**	**7[÷]**	**8**	**9**	**	or	1234	5	67	8	9	invalid	or
1	2345	6	7	89	or	1	2345	6	78	9	or
1	2345	67	8	9	or	All invalid					

1+	**2x**	**3456[x]**	**78[÷]**	**9**	*	or	1	2	3456	7	89	invalid	or
12	**3456x**	**7÷**	**8**	**9**	*	or						

1	2	3	4567	89	or	12	3	4567	8	9	or
1	23	4567	8	9	or	All invalid				

1÷	**2x[÷]**	**34[x]**	**5678**	**9**	*	or	1	23	4	5678	9	invalid	or
12÷[x]	**3[÷]**	**4**	**5678**	**9**	**	or						

1	2	3	45	6789	or	1	2	34	5	6789	or
1	23	4	5	6789	or	12	3	4	5	6789	**All invalid**

Applying BEDMAS to valid options in bold can give whole-number results, but only gives whole numbers on switching x and ÷ for two ** options. The * options must have ÷ or ÷,x combination, or alternatives [ ], as above. List ** options with +,- and -,+ alternatives.

1234+56÷7x8-9=1234+8x8-9=1234+64-9=1298-9= +1289   **Flush [Jack][Ace]289 ♠**
1234+56x7÷8-9=1234+392÷8-9=1234+49-9=1283-9= +1274   **Flush [Jack][Ace]274 ♥**

1234-56÷7x8+9=1234-8x8+9=1234-64+9=1170+9= +1179   Pair of Aces
1234-56x7÷8+9=1234-392÷8+9=1234-49+9=1185+9= +1194   Pair of Aces

12÷3x4+5678-9=4x4+5678-9=16+5678-9=5694-9= +5685   Pair of 5s
12x3÷4+5678-9=36÷4+5678-9=9+5678-9=5687-9= +5678   Flush [Jack]5678 ♥

12÷3x4-5678+9=4x4-5678+9=16-5678+9= -5662+9= -5653   <0 not whole-number
12x3÷4-5678+9=36÷4-5678+9=9-5678+9= -5669+9= -5660   <0 not whole-number

Of the higher value of a pair only +1289 can be displayed as a 'flush' e.g. [Jack][Ace][2] [8][9] spades (with other value +1274 e.g. [Jack][Ace][2][7][4] hearts).

7 SQUARE PHONE NUMBERS

Answer: 729

The first point to note is that there are only two three-digit perfect squares including the number 3 and in both, it is the most significant digit so 324 or 361 must be there.

A perfect square can only end in 0 (root ends in 0), 1 (root ends in 1 or 9), 4 (root ends on 2 or 8), 5 (root ends in 5), 6 (root ends in 4 or 6) or 9 (root ends in 3 or 7), and there are further restrictions in that

    a) those ending in 0 must have another 0 in front of it – ruled out
    b) those ending in 1, 4 or 9 must have an even number as the preceding digit.
    c) those ending in 5 must have a 2 as preceding digit.
    d) those ending in 6 must have an odd number as preceding digit.

As 2 3 7 and 8 are ruled out in the last place, it is likely that 7 and 8 will be the awkward customers. For 7, only 576, 729 and 784 are candidates as 676 is ruled out. The following pairs can thus co-exist:

324 and 576 leaves 1 8 9 – no good
361 and 729 leaves 4 5 8 – no good
361 and 784 leaves 2 5 9, and 529 qualifies

Thus Andrea is on 361, Bertha on 529 and Dorothy on 784 so that Caroline is between 576 and 729 i.e. 576, 625, 676 or 729 and Elizabeth is on 841, 900 or 961. The three already announced total 1674 so we will be adding anything between 1417 and 1690 for the other two. These would give from 3091 to 3374 for the grand total.

The only four-digit perfect squares lying in that bracket are 3136, 3249 and 3364, i.e. the squares of 56, 57 and 58 respectively.

The only qualifiers are 729 and 961 totalling 1690 to be added to 1674 for 3364.

Thus Caroline is on 729.

8   SOME PERMUTATIONS

Answer: 1, 2, 5, 8 and 9

Let the three digits be a, b and c.
As there are three digits there must be 3! (factorial 3) or 6 number of permutations for these digits, with 2 (or 3!/3 or (3-1)!) occurrences of a, b and c in each unit position. One permutation is the number 100a + 10b + c. With 2 occurrences of each number in each unit position the total of all permutations must be –
$$100(2a + 2b + 2c) + 10(2a + 2b + 2c) + (2a + 2b + 2c) = 111 \times 2 \times (a + b + c) \qquad (1)$$
It can be seen that for 'n' digits this generalizes to –
Total of all permutations of n digits = (111.... [n 1's]) × (n – 1)! × (sum of the n digits)   (2)

For three digits comprising all different numbers the total sum of digits must be a number between 6 (1,2,3) and 24 (7,8,9). The permutations total to 222 times the sum of digits, and the only sum of digits which gives three 3s is 15, with 3330 permutations. a + b + c must equal 15 so possible combinations of numbers are: (1,5,9), (1,6,8), (2,4,9), (2,5,8), (2,6,7), (3,4,8), (3,5,7) and (4,5,6)

Adding two further, different number digits, can only increase the 15 sum of digits by between 1 and 17. The following table shows, for n = 5 in equation (2), and sum of digits between 16 and 32, the sum of all 5 digit permutations.

Sum of 5 Digits	16	17	18	19	20	21
Total of permutations	4,266,624	4,533,288	4,799,952	5,066,616	5,333,280	5,599,944
Sum of 5 Digits	22	23	24	25	26	27
Total of permutations	5,866,608	6,133,272	6,399,936	6,666,600	6,933,264	7,199,928
Sum of 5 Digits	28	29	30	31	32	
Total of permutations	7,466,592	7,733,256	7,999,920	8,266,584	8,533,248	

Only for sum of digits equal to 25 are there five 6s in the total. The two additional digits must therefore add up to 10: (1,9), (2,8), (3,7) or (4,6). Combining this with the possible values of a,b,c gives:

Possible five digits	Product
1,2,5,8,9	720
1,2,6,7,9	756
1,3,4,8,9	864
1,3,5,7,9	945
1,3,6,7,8	1008
1,4,5,6,9	1080
2,3,4,7,9	1512
2,3,5,7,8	1680
2,4,5,6,8	1920
3,4,5,6,7	2520

The smallest possible product is 720. The five digits, in ascending order, are therefore 1, 2, 5, 8 and 9.

9 NERO SCRAG FROM CARREGNOS

Answer: 1,2,3,4,5,6,7

Mulling over given facts leads to eliminations as follows:-

Number of numerals N<10 and number of letters L<24 (Greek alphabet not all used)

Possible 4-digit combinations = N^4

Possible letter combinations = L^3

$10^7 <= (N^4) \times (L^3) < 10^8$ [8-figure total number of such vehicle registrations]

Max. L=23, so max L^3=12167, so min $N^4 > (10^7)/12167 > 800$, **so min N>5**

$(L^3 - N^4) > N^4$, so $L^3 > 2(N^4)$

[all 4-dig number patterns can be cherished, but no. of letter patterns that can't be > can]

If N=9, $2(9^4) = 2 \times 6561 = 13122 > 23^3[=12167]$, **so 5<N<9**

N=8, N^4=4096, so $L^3 > 2 \times 4096 > 8000$, **so L>20**

N=7, N^4=2401, so $L^3 > 2 \times 2401 > 4800$, **so L>16**, but L=17, 18 and 19 invalid, because 19^3=6859 means no 9999th, 8888th or 7777th letter triad, so at least 7, 8, 9 and 0 must be omitted, contradicting N=7 – **hence L>19**

N=6, N^4=1296, so $L^3 > 2 \times 1296 > 2500$, so L>13, but $L^3 > (10^7)/1296$, **so L>19**

Tabulate $(L^3 - N^4) > N^4$, so $L^3 > 2(N^4)$

	L	23	22	21	20
	[L^3]	[12167]	[10648]	[9261]	[8000]
N [N^4]					
8 [4096]		8071	6552	5165	$L^3 < 2(N^4)$
7 [2401]		9776	8247	6860	5599
6 [1296]		10871	9352	7965	6704

For L=23 and 22 – guessing L correctly, but N incorrectly leaves two possible options, both of which omit 0 and at least one other indeterminate numeral, because L^3>9999 so a 9999th letter triad exists and all numerals 1 to 9 can be included and hence 5, 6, 7, 8 or 9 omitted.

For L=21 – guessing L correctly, but N incorrectly leaves two possible options –

If N=8 guessed incorrectly, both other options omit 0 and 9 with at least one other indeterminate numeral, because 8888<L^3<9999, so an 8888th, but no 9999th letter triad exists and so only numerals 1 to 8 can be included with 5, 6, 7 or 8 taken for omissions.

If N=7 or 6 guessed incorrectly, then N=8 option omits just 0 and 9 definitely, because no 9999th letter triad, but other possibly correct option is indeterminate as above.

For L=20 – If N=7 guessed incorrectly, then N=6 must omit 0, 8, 9 and one indeterminate numeral, because 7777<L^3<8888 so 7777th, but no 8888th (or 9999th) letter triad exists and so only numerals 1 to 7 can be included with any taken to make up omissions.

If N=6 guessed incorrectly, then N=7 so only 0, 8 and 9 omitted, as above.

So, the numerals used are 1 to 7.

10 PLANTATION

Answer: 4 yards

Consider small plots, e.g. with 4, 8 and 12 trees.

If the trees are d yards apart this would give areas of d^2, $4d^2$ and $9d^2$ respectively.

Thus 4n trees give an area of $n^2.d^2$

number of acres = number of trees

$n^2.d^2 / 4840 = 4n$

$n = 4.4840 / d^2$

$n = 2^5.5.11^2 / d^2$

Consider possible values for d and the resulting values for n and the number of trees (4n) –

d	n	4n
1	19360	77440
2	4840	19360
4	1210	4840
11	160	640
22	40	160
44	10	40

The 'given' digit must be 8, ie there must be 4840 trees.

The trees would be **4 yards** apart.

11 POWER STRUGGLE

Answer: 32775

Let my number be X, the nearest power of 2 be 2^M, and the nearest perfect square be N^2.

(a) The distance from X to N^2 must be double that from X to 2^M.
[Otherwise X will be an even distance from 2^M and cannot be odd.]

(b) M will be odd.
[Otherwise the power of 2 will itself be a perfect square closer to X than N^2.]

(c) N will be odd.
[Because N^2 is an even distance from the odd number X.]

(d) X cannot be between N^2 and 2^M.
[If it were, then the distance between N^2 and 2^M would be divisible by 3. But N^2 has remainder 0 or 1 when divided by 3, and 2^M (with M odd) has remainder 2 when divided by 3. So the two numbers cannot differ by a multiple of 3.]

So we have one of the following two cases (for some whole number d):

	Case A				Case B		
N^2	2^M	X	$(N+1)^2$	$(N-1)^2$	X	2^M	N^2
Distances: d	d	e(>2d)		e(>2d)	d	d	

Note that N^2 is also the nearest square to 2^M and, given 2^M, then N is easily found using a calculator. We tabulate the possibilities:

M	2^M	N	N odd?	N^2	Case	d	X	e	e>2d?
3	8	3	yes	9	B	1	7	3	yes
5	32	6	no						
7	128	11	yes	121	A	7	135	9	no
9	512	23	yes	529	B	17	495	11	no
11	2048	45	yes	2025	A	23	2071	45	no
13	8192	91	yes	8281	B	89	8103	3	no
15	32768	181	yes	32761	A	7	**32775**	349	yes
17	131072	362	no						
19	524288	724	no						

So the only two odd numbers less than a million that work are 7 (given in the question) and 32775.

Answer: Catter, Eighties, Forkpoynt and Hegroom

Firstly, consider all the elements of the forecast, and deduce from them which regions could possibly be next to which others. This leads to the following possibilities, where for brevity the initial letters of the regions are used.

A could neighbour:- B,C,D,E,F,G,H. B:- A,C,D,F,G,I. C:- A,B,E,F,I. D:- A,B,E,F,G,H. E:- A,C,D,H,I. F:- A,B,C,D,H,I. G:- A,B,D,H,I. H:- A,D,E,F,G,I. I:- B,C,E,F,G,H.

Now, the 9 regions are on a map on a plane, with every region touching exactly 4 others, and with a ring around the island formed by all the coastal regions A, D, F, H (in some order). There are only 3 essentially different orders for the coastal regions, because starting the order at different points, or reversing the order, won't alter the way regions neighbour each other. It is easier to sketch diagrams with points representing the regions and lines between the points representing regions being adjacent; between any two points there is at most one line, and no lines can cross.

A straightforward solution method would be to apply orderly trial and error from here, progressively considering possible neighbouring regions; but that is very laborious. It is better to concentrate first on what overall patterns would be legal. A number of local configurations – as follows in Figures 1 to 4 – can't have lines added in the plane to give 4 lines to every point.

No coastal region can touch 3 other coastals (Figure 1), and so each must have 2 coastal neighbours, and therefore 2 offshore neighbours; so that leaves 4 x 2 = 8 'adjacencies' to be between a coastal and an offshore region. No offshore region can touch 3 coastals (Figure 2), and no two offshore regions can touch the same two coastals (Figure 3). So the 5 offshores touch 2,2,2,1,1 or 2,2,2,2,0 coastals. The first of these alternatives would have the regions arranged like Figure 4, and you will find that again you can't add edges to this on a plane so that every point has 4 lines. The second alternative has 4 offshores touching the coastals like Figure 5, and the 5th offshore must then touch all the other 4 offshores; and, looking at the list of possible neighbourings, this 5th offshore can only be I.

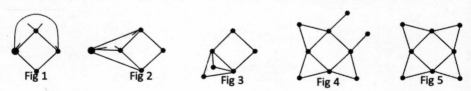

Fig 1   Fig 2   Fig 3   Fig 4   Fig 5

So every pair of adjacent coastals has an offshore neighbouring both of them. But if the coastals F&H were adjacent, the list of possible neighbourings shows the only offshore that could neighbour the pair of them is I; but this doesn't work, because I has just been proved to be needed elsewhere. So the circular order of coastals must be A,F,D,H (or the reverse). B is the only candidate for touching both F&D, and that leaves C as the only available candidate for touching both A&F. Either of E or G could touch either of the pairs D&H or H&A.

Now, to complete the planar diagram with every point having 4 lines, the 4 offshores that touch coastals must have one more line each. If the circular order is C,B,E,G this is impossible because there is only one possible neighbouring, namely C with B. But if the circular order is C,B,G,E then E can neighbour C, and B can neighbour G, which gives a legal solution to the puzzle.

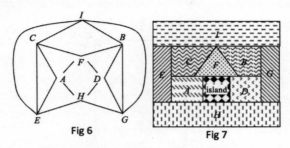

Fig 6

Fig 7

Therefore the solution to the question posed is that the regions neighbouring A are C,E,F&H. A possible diagram is Figure 6. (Actually, without the labellings, the pattern of interconnections between the points is the only one possible on a plane with 9 points and 4 lines at every point.) Going back to an 'ordinary' type of map, a possible disposition of the regions is Figure 7.

Answer: 17 in.

Total area of shield (L=top width)=Square Area+Semi-circle Area=$L^2+\pi(L^2)/8$
BF bisects square ABDF, so Area of ABF=$(L^2)/2$=Area of lower white zone 'CDEC'

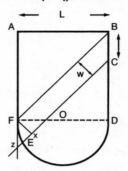

So Area in white=L^2 and so black band area=$\pi(L^2)/8$

FB=L√2   **Fz=v=w√2**   **zx=Fx=w**

By visual inspection: Area of [BCxF]<Area of band BCEF<Area of BCzF

Area BCzF=(FB)(Fx)=wL√2=vL>$\pi(L^2)/8$; **hence v/L>π/8>0.39**

Area [BCxF]=Area BCzF-Area Fxz=(vL)-[(Fz)^2]/4

=vL-(v^2)/4<$\pi(L^2)/8$, so **8(v/L)-2(v/L)^2<π and so v/L<0.45**

So 0.39<v/L<0.45 is approx. PxQ/(N^3) with P, Q (both primes) and N all odd, 9<PxQ<100
N=3, no valid PxQ gives v/L in range
N=5, only PxQ=51 gives v/L in range; 51/125=0.408 [=41% rounded]
N>=7, no valid PxQ gives v/L in range

Two other shields have L<24 in. and v/L=41% rounded. Test for v/L in range [0.405, 0.415]

L=	2	3	4	5	6	7	8	9	10	11
v<L/2	1	1	1	1,2	1,2	1,2,3	1,2,3	1 to 4	1 to 4	1 to 5
v/L						all invalid				

L=	12	13	14	15	16	17
v<L/2	1 to 5	1 to 6	1 to 6	1 to 7	1 to 7	1 to 8
v/L	5/12=0.416.. close, but this and all the above invalid					**7/17=0.411...**

L=	18	19	20	21	22	23
v<L/2	1 to 8	1 to 9	1 to 9	1 to10	1 to10	1 to11
v/L			all the above invalid		**9/22=0.409...**	all invalid

Only 51cm/125cm=0.408; 7in/17in=0.411..; and 9in/22in=0.409.. round to 41%
With 125cm>1m>36in. the shortest L is 17 in.

14 ENDLESS NUMBER

Answer: 381654729

It is clear that when we are down to five digits, the number must divisible by 5. So the fifth digit is five. Similarly, if the 2nd, 4th, 6th and 8th numbers are divisible by the appropriate number, the least significant number must be even to have any chance at all. Thus, we know that the nine digits must be ODD EVEN ODD EVEN 5 EVEN ODD EVEN ODD.

First considering the evens, if a number is to be divisible by two, its second digit must be even so that the following are candidates:
12 14 16 18 32 34 36 38 72 74 76 78 92 94 96 98

If a number is divisible by 4, the last two digits must be a multiple of 4, so that, noting that 5 is already accounted for, the following are candidates for the third and fourth digits: 12 16 32 36 72 76 92 96; we have established that the fourth digit must be 2 or 6.

For a number to be divisible by 6, the sum of its digits must be divisible by 3. We know that the fifth digit must be 5.
For a number to be divisible by 8, the last three digits must be a multiple of 8; thus the following are candidates for digits 6, 7 and 8:
216 296 416 432 472 496 632 672 816 832 872 896

With 2 and 6 already accounted for above, numbers including both are ruled out; we are left with 416 432 472 496 816 832 896. Thus we know that the eighth digit is 2 or 6 and thus both can be ruled out of the other positions. Furthermore, we now know that the second and sixth digits must be 4 or 8.

For a number to be divisible by 6, the sum of its digits must be divisible by 3. We know that the fifth digit must be 5 and the sixth even. So, the fifth and sixth digits must be one of the following: 52 54 56 58 but 2 and 6 are already ruled so we are left with 54 or 58.

Now we try to link up possible candidates, remembering that 2 and 6 can only be in the fourth or eighth digits and 4 and 8 can only be in the second and sixth digits.

Starting with 2 and 4: 14 32, 14 36, 14 72, 14 76, 14 92, 14 96, 18 32, 18 36, 18 72, 18 76, 18 92, 18 96, 34 12, 34 16, 34 72, 34 76, 34 92, 34 96, 38 12, 38 16, 38 72, 38 76, 38 92, 38 96, 74 12, 74 16, 74 32, 74 36, 74 92, 74 96, 78 12, 78 16, 78 32, 78 36, 78 92, 78 96, 94 12, 94 16, 94 32, 94 36, 94 72, 94 76, 98 12, 98 16, 98 32, 98 36, 98 72, 98 76

A long list but we can quickly eliminate a great deal by noting that if a number is to be divisible by 3, the sum of its digits must also be divisible by 3. So we look at the totals of the first three digits in the above quartets and we are left with: 14 72, 14 76, 18 32, 18 36, 18 92, 18 96, 38 12, 38 16, 38 72, 38 76, 74 12, 74 16, 78 32, 78 36, 78 92, 78 96, 94 72, 94 76, 98 12, 98 16, 98 72, 98 76

Now we add the total of these quartets to 5 and see if by adding 4 or 8, the required sixth digit, if available, we can get to a multiple of 3. We are left with: 14 72 58, 14 96 58, 18 36 54, 38 16 54, 38 76 54, 74 12 58, 78 36 54, 78 96 54, 98 16 54, 98 76 54

We now try to tie up the displayed sixth digit with the three-digit multiples of 8: 216 296 416 432 472 496 632 672 816 832 872 896

We are left with

14 72 58 96 3
14 96 58 32 9
14 96 58 72 3
18 36 54 72 9
38 16 54 72 9
74 12 58 96 3
78 96 54 32 1
98 16 54 32 7
98 16 54 72 1

Finally, we need the first seven digits to be divisible by 7. 3816547/7 = 545221; the others all go to fractions. Thus the phone number is 381654729.

15 DISCS A GO-GO

Answer: 630mm and 210mm

Let 'r' be the radius of the disc, 't' be the length of each of the triangle's sides and 'h' be the length of each of the regular hexagon's sides. If the probability of the disc covering more than one tile is the same for both shapes then so must the probability of it not doing so be the same for both shapes. For it to fall completely within one tile then the center of the disc must fall within an area within each tile at least one radius width from the edge. Such an area would cover a similar shape within each tile to that of the tile itself, and the probability of only covering one tile would be the ratio of this inner area to the whole area. This ratio is the same as the square of the ratio of the respective sides (ie. $(ab)^2 : (AB)^2$ below).

Triangular Tiles

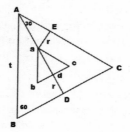

Hexagonal Tiles

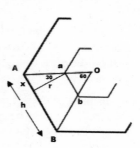

In $\triangle$ AaE   Aa = r / sin 30 = 2r
In $\triangle$ ABD   AD = t cos 30 = t$\sqrt{3}$/2
AD = Aa + ad + dD
t$\sqrt{3}$/2 = 2r + ab$\sqrt{3}$/2 + r
ab = 2(t$\sqrt{3}$/2 – 3r) /$\sqrt{3}$ from which –
$(ab)^2/(AB)^2 = (t – 2r\sqrt{3})^2 * 1/t^2$ (i)

ab = AB – 2x
x = r * tan 30 = r($\sqrt{3}$/3)
ab = h – 2r($\sqrt{3}$/3)
$(ab)^2/(AB)^2 = (h - 2r\sqrt{3}/3)^2 * 1/h^2$ (ii)

Substituting t = 3h in equation (i) it becomes after simplification the same as equation (ii). For the probabilities to be the same equation (i) must equal equation (ii), which is therefore when t = 3h.

't' and 'h' are two triangular numbers of less than four digits such that t = 3h.

The triangular numbers less than 1000 are easily worked out as follows:
1 3 6 10 15 21 28 36 45 55 66 78 91 105 120 136 153 171 190 210 231 253 276 300 325 351 378 406 435 465 496 528 561 595 630 666 703 741 780 820 861 903 946 990

The only pair of these where one is three times the other and both are even is 210 and 630, so t = 630mm and h = 210mm.

16 PROCEED TO CHECKOUT

Answer: 1, 5 & 22

The method of solution depends largely on organised trial and error, as suggested by the pub's name. Simple arithmetic is all that is needed, and there are some observations that simplify the process.

However many sectors there are, the smallest label has to be 1, or otherwise the Lowest Impossible Checkout (call it LIC for brevity) would be 1 itself.

Now, given that the first sector is labelled 1, many values can be ruled out for the second or third sector, because they would allow 36 as a possible checkout. These include: 4, because 3 of 3x4 = 36; 6, allowing 2 of 3x6; 7, allowing 3x7 + 2x7 + 1x1; 9, allowing 3x9 + 1x9; 10, allowing 3x10 + 2 of 3x1; 11, allowing 3x11 + 3x1; 12, allowing 3x12; 15, allowing 2x15 + 2 of 3x1; 16, allowing 2x16 + 2 of 2x1; 17, allowing 2x17 + 2x1; and 18, allowing 2x18.

Now, with one sector, with the label 1, the possible checkouts are: 1, 2 & 3, which can be done using one dart; 4, 5 & 6, which can be done using two darts but no fewer; and 7, 8 & 9, which can be done using three darts but no fewer. The LIC is 10 – this would need at least four darts.

So the second smallest label can't exceed 10 (or otherwise the LIC would still be 10) and the second smallest label must be at least 1 more than the smallest label, i.e. at least 2. Then, because of the ruling out of values as noted above, the second smallest label can only be 2, 3, 5 or 8.

Now analyse combinations of three sectors, with first sector 1 and second sector each of these four candidate labels in turn.

{1,2,5} allows 2 of 3x5 + 3x2 = 36 as a possible checkout; similarly {1,2,8} allows 2 of 2x8 + 2x2; {1,2,12} allows 3x12; {1,2,13} allows 2x13 + 3x2 + 2x2; and {1,2,14} allows 2x14 + 2 of 2x2. You will find that {1,2,3} has 23 as an impossible checkout, and {1,2,n} for n≥18 has 17 as an impossible checkout, both of which of course are lower than 36. This has covered all possibilities for third sector labels when the second sector is labelled 2, and none of them gives 36 as the LIC.

{1,3,5} allows 2 of 3x5 + 2x3 = 36; {1,3,8} allows 3x8 + 3x3 + 1x3; {1,3,13} allows 2x13 + 3x3 + 1x1; {1,3,14} allows 2x14 + 2x3 + 2x1; and {1,3,21}allows 1x21 + 3x3 + 2x3.

You will find that {1,3,19} has 32 as impossible; {1,3,20} has 33 as impossible; {1,3,22} has 35 as impossible; {1,3,n} for n≥23 has 22 as impossible; and all of these are lower than 36. This exhausts all third sector possibilities, with none of them giving 36 as the LIC.

{1,5,8} allows 3x8 + 2x5 + 2x1 = 36; {1,5,13} allows 2x14 + 1x5 + 3x1; {1,5,14} allows 2x14 + 1x5 + 3x1; {1,5,19} allows 1x19 + 3x5 + 2x1; {1,5,20} allows 1x20 + 3x5 + 1x1; {1,5,21} allows 1x21 + 3x5; {1,5,23} allows 1x23 + 2x5 + 3x11; {1,5,24} allows 1x24 + 2x5 + 2x1.

You will find that {1,5,n} for $n \geq 25$ has 24 as impossible, which is less than 36. This exhausts all third sector possibilities except {1,5,22}, which you will find indeed does have 36 as its LIC.

{1,8,13} allows 2x13 + 1x8 + 2x1 = 36; and {1,8,14} allows 2x14 + 1x8. You will find that {1,8,n} for $n \geq 16$ has 15 as impossible, which is less than 36. So this has exhausted all possibilities with the second sector labelled 8.

Therefore the only possibility satisfying the Teaser's condition that the Lowest Impossible Checkout must be 36 is for the three sectors to be labelled 1, 5 & 22.

17 ROSE GARDEN

Answer: 101 and 146

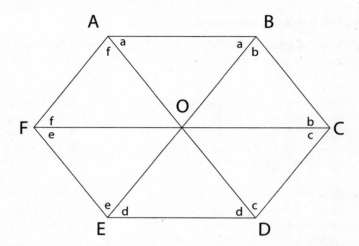

A hexagon in a circle consists of six isosceles triangles, as shown above, where O is the centre of the circle and OA = OB = ... = OF. The sum of all the angles in the six triangles is 6x180 = 1080. The sum of all the angles at the centre of the circle is 360, so the remaining angles add up to 1080-360. Therefore, 2a+2b+2c+2d+2e+2f = 720, and a+b+c+d+e+f = 360.

The internal angles at A, C and E add to (f+a) + (b+c) + (d+e), i.e. a+b+c+d+e+f.
Therefore A+C+E = 360. Similarly, B+D+F = 360, so we are looking for two three-digit triads, each adding to 360. It is evident that the repeated digit must be 1.

The 'units' of one triad must total 20 while its 'tens' must total 4(0). This gives a digit total of 24.
The 'units of the other triad must total 10; its 'tens' must total 5(0). This gives a digit total of 15.
The digit total for the twelve digits of the six angles is hence 39 (neglecting leading ones). This includes four ones and eight other different digits. If all of the digits from 0 to 9 were included, the sum would be 0+1+1+1+1+2+3+4+5+6+7+8+9 = 48
It follows that the missing digit is 9.

The only possibility for a units total of 20 is 8, 7, 5.
Consider a tens total of 4(0); this suggests either 1, 1, 2 or 0, 1, 3.
1, 1, 2 leads to either 0, 1, 4 or 1, 1, 3 for the tens total of 5(0).
However, 0, 1, 3 can be neglected as it cannot lead to any options for a tens total of 5(0).

We therefore have three possibilities for one of the triads-

125, 118, 117
127, 118, 115
128, 117, 115

We have nine possibilities for the other triad-

101, 113, 146
101, 116, 143
103, 111, 146
103, 116, 141
106, 111, 143
106, 113, 141
110, 114, 136
110, 116, 134
114, 116, 130

In order to have a perfect power and two primes we must have 125, 118, 117 with 101, 113, 146, or 128, 117, 115 with 101, 113, 146. In either case, the smallest and largest angles are 101 and 146.

18 CUT FOR PARTNERS

Answer: Ace of clubs and nine of spades

We number the cards ♣2 = 1, ♦2 = 2, etc. up to and including ♥A = 51 and ♠A = 52. Then, after two cards have been drawn, there are fifty still available. Assuming that George drew card X and Martha drew card Y, the chance that the visitors will partner each other is worked out by the chance that they both draw cards higher than X or both draw cards lower than Y. Thus:

P = ((52 - X)!/((50 - X)!2!) + ((Y - 1)!/(Y - 3)!)/2!)/(50x49)= ((52 - X)(51 - X) + (Y - 1)(Y - 2))/4900   (1)

We are given that P/2 = the above equation with ♥7 = 23 in for Y   (2)
and that P/3 = the above equation with ♥6 = 19 in for Y   (3)

We thus have three simultaneous equations with three unknowns P, X, Y. We solve as follows:

Divide equation (2) by equation (3) with ♥7 = 23 and ♥6 = 19:

1.5 = ((52 - X) x (51 - X) + 462)/((52 - X) x (51 - X) + 306)

So 1.5 x (52 - X) x (51 - X) + 459 = (52 - X) x (51 - X) + 462

So $1.5X^2$ - 1.5 x 103X + 1.5 x 2652 - X^2 + 103X - 2652 - 3 = 0

So $0.5X^2$ - 51.5X + 1323 = 0

So X^2 - 103X + 2646 = 0

That gives X = (103 +/ - $(10609 - 4 x 2646)^{0.5}$)/2

gives X = (103 + / - 5)/2 = 49 or 54

However, 54 is out as the maximum is 52 so 49 i.e. the ace of clubs, is correct.

Now divide equation (1) by equation 2:

2 = (6 + (Y - 1)(Y - 2))/(6 + 22 x 21)
2 x 468 = 6 + Y^2 - 3Y + 2

So that Y^2 - 3Y - 928 = 0

Gives Y = (3 +/ - $(9 + 4 x 928)^{0.5}$)/2
= (3 + / - 61)/2
= (must be positive) 32 i.e. the nine of spades.

19 TOTAL RESISTANCE

Answer: 22 pupils, 1052 ohms

There are 22 possible resistances using five resistors of resistance r ohms; this can be demonstrated by an exhaustive search. For each of R=2r and R=r/2, there are two possible networks giving that resistance (but only one is shown):-

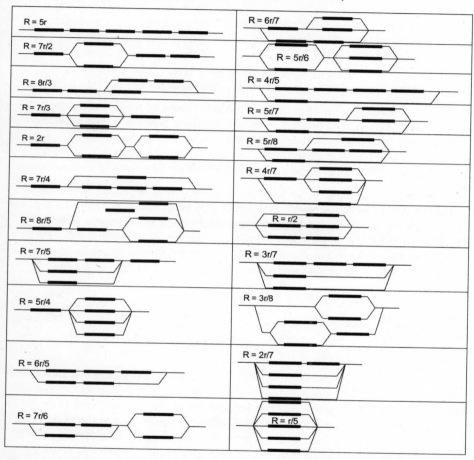

The total resistance of these 22 networks is r*(5 + 7/2 + 8/3 + 7/3 + 2 + 7/4 + 8/5 + 7/5 + 5/4 + 6/5 + 7/6 + 6/7 + 5/6 + 4/5 + 5/7 + 5/8 + 4/7 + 1/2 + 3/7 + 3/8 + 2/7 + 1/5) = r*(1052/35). As r = 35 ohms the total resistance is 1052 ohms.

Answer: 7, 7 & 4 blocks

The problem is obviously to find out the relative positions of the hotel, diner, library, museum and gallery on a square grid, where the distances are the number of edges of the grid. The method of solution is to find possible triplets of positions for hotel/diner/library with distances 9/5/6, and for hotel/diner/museum with distances 9/4/7, and hotel/diner/gallery with distances 9/3/8. Each distance will be fulfilled by a certain number of edges in one orthogonal direction plus a number in a perpendicular direction; the conditions of the teaser prohibit any distance being in a straight line. (Note that the points a distance n from a given point form a square rotated by 45°, and the vertices will be excluded because of that last condition.)

The 9-block distance between the hotel and the diner could be fulfilled as 8+1 or 7+2 or 6+3 or 5+4. Then you will find that each of these could give one or two positions for each of the library, museum and gallery. (The situation is analogous to two circles intersecting, which would in general be in two points if they were close enough, but sometimes the prohibition of straight-line distances will eliminate one of the possibilities.)

Positions for the various places can be specified using (x,y) co-ordinates, where the x co-ordinate could signify the North, East, South or West direction, and the y co-ordinate could signify the perpendicular direction either clockwise or anti-clockwise from the x direction. Without loss of generality the hotel can be put at $(0,0)$, with the diner's co-ordinates both being positive and having $x{\geq}y$. Then possible positions for library, museum and gallery correspond to the diner's position as in the following table.

Diner	(8,1)		(7,2)		(6,3)		(5,4)	
Library	(4,2)	(5,-1)	(3,3)	(5,-1)	(2,4)	(5,-1)	(1,5)	
Museum	(5,2)	(6,-1)	(4,3)	(6,-1)	(3,4)		(6,1)	(2,5)
Gallery	(6,2)	(7,-1)	(5,3)		(7,1)	(4,4)	(6,2)	(3,5)

It now remains to identify positions for the library, museum and gallery that correspond to the same diner position but don't have any x co-ordinates or y co-ordinates in common. With the diner at (8,1) there are only two possible choices for y co-ordinates (2 or -1), so three places can't have their y co-ordinates all different. Similarly when the diner is at (7,2) (y co-ordinates being 3 or -1). If the diner is at (6,3) then the museum must be at (3,4), and to avoid having another place's y co-ordinate being 4 as well this forces the library to be at (5,-1) and the gallery to be at (7,1); but these do not conflict, so this is a compliant solution. Lastly, if the diner is at (5,4) then the other places must have either a y co-ordinate of 5 or an x co-ordinate of 6, so the three places can't avoid conflicting.

So the unique solution is
hotel at (0,0),
diner at (6,3),
library at (5,-1),
museum at (3,4), and gallery at (7,1).
This means that the required distances
are 7, 7 & 4 blocks.

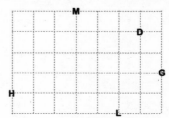

21 EVEN OR ODD

Answer: 315

For the process of successive steps eventually to lead to 1, it must at some earlier stage reach a power of 2. That power of 2 must equal $13n + 1$, for some integer n, so we test the following integers for divisibility by 13: 3, 7, 15, 31, 63, 127, 255, 511, 1023, 2047, 4095....

We find that the first number in this sequence divisible by 13 is 4095 = 13 times 315. So starting with 315, the game leads to 1 in 13 steps. If we start with any other odd starting number, its sequence must take more steps to reach 1, if at all, as it must pass through a power of 2 not less than 4096.

Answer: 11

In the 1st pile there are B blacks and 3B reds.
In the 2nd pile there are 2C - 1 blacks and C reds.
In the 3rd pile there are D blacks and ND - 1 reds, where N > 1.

Hence B + 2C - 1 + D = 26 and 3B + C + ND - 1 = 26

Subtracting the first equation from twice the second,
27 - 5B = (2N - 1)D
C = (27 - B - D)/2

The four possibilities are tabulated below:

							Piles	
B	27-5B	2N-1	N	D	C	1	2	3
1	22	11	6	2	12	4	35	13
2	17	17	9	1	12	8	35	9
3	12	3	2	4	10	12	**29**	11
4	7	7	4	1	11	16	32	4

Liam must have told me how many cards were in the second pile as prime numbers are present in two possibilities for the third pile.

There were **11 cards** initially in the third pile.

23 FAMILY STOCK

Answer: 1, 2, 3 and 6

Let the grandchildren's ages be a, b, c and d.

At each's first birthday the number of shares transferred would be 1, at their second 3, third 5, etc. This is an arithmetic progression so that the nth term (a_n) is $a + (n-1)d$ and the sum of the first n terms (S_n) is $n\{2a + (n-1)d\}/2$. Where a is the first term (one), and d is the difference (two).

So $a_n = 1 + 2(n-1) = 2n-1$ and $S_n = 2\{2 + 2(n-1)\}/2 = n^2$

The total number of shares given over their lifetimes is therefore $a^2 + b^2 + c^2 + d^2$ and the percentage of this total which they received at their last birthday was $100(2a-1)/(a^2 + b^2 + c^2 + d^2)$, $100(2b-1)/(a^2 + b^2 + c^2 + d^2)$ etc. As these are all exact percentage then $(a^2 + b^2 + c^2 + d^2)$ must be a factor of 100 multiplied by any common factor of $(2a-1)$, $(2b-1)$, $(2c-1)$ and $(2d-1)$.

Factors of 100 are 1, 2, 5, 10, 20, 25, 50 & 100, and as $(2n-1)$ is an odd number a common factor is 1 with the possibility of it also being 3, 5, 7, etc.

For the lowest 3 common factors the tables below show the first few values of n, $(2n-1)$ and n^2. Note that only total values of n^2 equal to or less than 100 times each common factor or divisor need be considered.

2n-1 divisible by 1		
n	2n-1	n2
1	1	1
2	3	4
3	5	9
4	7	16
5	9	25
6	11	36
7	13	49
8	15	64
9	17	81

Sum of first 4 n2	30

2n-1 divisible by 3		
n	2n-1	n2
2	3	4
5	9	25
8	15	64
11	21	121
14	27	196
17	33	289

Sum of first 4 n2	214

2n-1 divisible by 5		
n	2n-1	n2
3	5	9
8	15	64
13	25	169
18	35	324

Sum of first 4 n2	566

For solution total of four different n2 must equal 50 or 100	Total of four different n2 must equal 300	No solutions as max total exceeds 5 * 100

In the table of 2n-1 divisible by 5 the lowest four values of different n^2 total 566. As this is greater than 100 * 5 there can be no solutions for common factors of 5. This is also the case for common factors of higher values. Therefore there can only be solutions from the above first two blocks.

The only four separate values of n^2 (the grandchildren are all of different ages) which add up to a factor of hundred multiplied by a common divisibility factor is from block one for ages 1, 2, 3 and 6 (n^2 for which total 50).

Answer: 8 & 5 centimetres, and 16; 7; 4 & 17; 10

The surface area of the pie is 20 x 13 = 260cm², so one fifth of this is 52cm². To work out the areas of the individual slices, the formula for the area of a triangle being half base times height is the best to use.

Let the internal point be p cm from the left side and q from the top. Because the point is internal and in the top left-hand quarter, $0 < p \le 10$ and $0 < q \le 6$. Let the distances clockwise from successive corners be w, x, y & z cm, with one of w', x', y' or z' cm as appropriate denoting the second distance on the side that has two cuts reaching it.

Looking first at the side that has two cuts reaching it, the shape of the resulting slice is a simple triangle and half the base times the height must be 52cm², so the base times the height must be 104 = 2^3 x 13. Given the dimensions of the pie, this means that either: the length along an edge must be 13 and the distance of the internal point from that edge must be 8; or vice versa. So the only possibilities are $p = 7$ (so $a - p = 13$) with the right-hand (x side) having two cuts meeting it 8cm apart, or $q = 5$ (so $b - q = 8$) with the bottom (y side) having two cuts meeting it 13cm apart.

The shape of each of the other slices is a quadrilateral incorporating a corner of the whole pie, and can be thought of as being composed of two triangles that share a side running from the internal point to that corner. This gives a series of integer equations that need to be solved.

For the $p = 7$ case, where the x side has two cuts reaching it, the distance from the second cut to the bottom right-hand corner must equal $13 - x - 8$, i.e. $5 - x$. Double the area of the slice enclosing the top right-hand corner would be $q(20-w) + 13x$, and for the one enclosing the bottom right-hand corner it would be $qy + 13(5-x)$, and these must both equal 104. Considering divisibility by 13, and the fact that $q < 6$, we must have $20 - w = 13$ and $y = 13$. Then adding both doubled areas and dividing by 13 gives the equation $2q + 5 = 16$, which is impossible.

For the $q = 5$ case, where the y side has two cuts reaching it, the distance from the second cut to the bottom left-hand corner must equal $20 - y - 13$, i.e. $7 - y$. So $0 < y < 7$. The four expressions derived from considering the doubled areas, taken clockwise from the bottom left-hand corner, are therefore $8(7-y) + pz$, $5w + p(13-z)$, $5(20-w) + (20-p)x$, and $8y + (20-p)(13-x)$, and these must all equal 104. Adding the first and second of these gives $8(7-y) + 5w + 13p$, which must equal 208. Considering divisibility by 13, and the limits on y and w, we must have $w = 7 - y + 13k$, where $k \ge 0$; this leads to the equation $p + 5k = y + 9$. With the limits of $0 < p \le 10$ and $0 < y < 7$, this has a number of possible solutions as shown below. Then ensuring $0 < w < 20$, and then requiring $z = (104 - 8(7-y)) / p = 8(y+6)/p$ to be a whole number, leaves only one possibility, and this is confirmed as a solution by checking that $x = (104 - 5(20-w)) / (20-p) = (5w+4) / (20-p)$ is also an integer. Finally, because the two cuts reach the y side 13cm apart, $y' = y + 13$.

y	1	1	2	2	3	3	**4**	4	5	5	6	6
k	0	1	1	2	1	2	**1**	2	1	2	1	2
p	10	5	6	1	7	2	**8**	3	9	4	10	5
w	6	19	18	~~31~~	17	~~30~~	**16**	~~29~~	15	~~28~~	14	~~27~~
z	~~56/10~~	~~56/5~~	64/6		~~72/7~~		**10**		88/9		~~96/10~~	
x							**7**					
y′							**17**					

(Incidentally, this is still the only solution if cuts are allowed to go exactly to one corner or more, but for brevity of the puzzle statement and its worked solution this loosening of the conditions was excluded.)

25 TURNIP PRIZE

Answer: 37

Each tile that was removed did not change the average area, so each such tile must have had an area equal to the previous average. In particular the average must be a whole number. Building up from 1, adding 2(1+2), then adding 3(1+2+3) etc shows that this happens first when the age is 10, then 13, etc. However, we'll use an algebraic approach rather than this long-winded numerical one.

Let his age be N. If we wanted to count the total of <u>all</u> products X.Y where X can range from 1 to N and Y can also range from 1 to N, (so that, for example 3.4 and 4.3 are both counted) then the total would be $(1 + 2 + ... + N)^2 = \frac{1}{4} N^2 (N + 1)^2$ = A, say. But if X and Y are different in the range 1 to N, then there is just one tile with those dimensions. Now A includes all the squares from 1^2 to N^2 (totalling $\frac{1}{6} N(N + 1)(2N +1)$ = B, say) and each other product (e.g. 3.4 and 4.3) twice. So the total area of the rectangles/squares in this puzzle is
$\frac{1}{2} (A - B) + B = \frac{1}{2} (A + B) = \frac{1}{24} [3N^2 (N + 1)^2 + 2N(N + 1)(2N +1)]$
This tidies up to **total area of tiles = $\frac{1}{24}$ N (N + 1) (N + 2)(3N + 1)**
The tiles consist of:
N with shorter (or equal shortest) side 1;
N-1 with shorter side 2;
N-2 with shorter side 3;
and so on to
1 with shorter side N.
So there are $\frac{1}{2}$ N (N + 1) tiles, making **the average area = $\frac{1}{12}$ (N + 2)(3N + 1)**

For $\frac{1}{12}$ (N + 2)(3N + 1) to be a whole number we need N+2 to be divisible by 3 (i.e. realistically N=10, 13, 16, ... , 49). If N is odd we also need 3N + 1 to be divisible by 4, (leaving just the odds 13, 25, 37 and 49) and if N is even we need N+2 to be divisible by 4 (leaving just the evens 10, 22, 34 and 46). We tabulate the results below:

N	Average area	Tiles of area equal to average
13	5.10	5x10
25	9.19	9x19
37	13.28	13x28 and 14x26
49	17.37	17x37
10	1.31	none
22	2.67	none
34	3.103	none
46	4.139	none

The two tiles that were removed had area equal to the average, and so **N=37**.

26 MEMBERS OF THE JURY

Answer: 41

Let the youngest age of the original jury be y and the common difference be a.
Then the ages of the original jury are y, y + a, y + 2a, etc., up to and including y + 11a with one of those appearing twice and one missing.

When the average age reduces from one prime number to another, that implies a drop of at least two so that the total age must reduce by 24, 48, 72, etc. and obviously 24 and 48 are the only serious candidates because of the age restriction. Looking at the candidate prime numbers for the average age, those in the range which have any chance at all, we have 29, 31, 37, 41, 43, 47, 53, 59.
Total age in each case will be 348, 372, 444, 492, 516, 564, 636, 708 respectively.
A drop of six will imply a total age change of 72 (impossible) so we are looking at twos or fours and the only practical possibilities are 47 to 43, 43 to 41 or 41 to 37.
A drop of four implies a total age drop of 48 years which is impossible as 65 - 20 = 45. So it has to be 43 dropping to 41.
We have 43 x 12 = 516 for the total age of the original jury to be reduced to 41 x 12 = 492 for the replacement jury.
Consider possible values of a. If a = 4 or more, the age spread will be at least 48, clearly too much.
If a = 1, the spread will be 12, clearly too little. So a = 2 or 3.
Starting with a = 2, we can have anything from 20 to 42 to inclusive up to 43 to 65 inclusive. However, the drop of 24 in the reduction rules this out.
So a = 3, we can have anything from 20-53 to 32-65. We need approx. 516/6 = 86 to have any chance so it will be something like 23-56, 26-59, 29-62 or 32-65. Intermediates are ruled out as they will hit perfect squares when three is added, viz. 36 and 49.
20-53 totals 6 x 73 = 438, hopelessly inadequate.
Starting with 23-56, the original jury is 23 26 29 32 35 38 41 44 47 50 53 56 with one number over 40 duplicated and one missing. The total is 474 and we need 516, i.e. an extra 42. An extra 56 (maximum in this range) implies an omission of 14, which is impossible.
26-59 implies an original jury of 26 29 32 35 38 41 44 47 50 53 56 59 with one omission and one number over 40 repeated. The total is 510 and we need 516, i.e. an extra 6. Legal possibilities are;
26 29 32 38 41 41 44 47 50 53 56 59
In the replacement jury, 35 is needed for the series so the only legal possibility would be 35 in 59 out.
That would give
26 29 32 35 38 41 41 44 47 50 53 56
26 29 32 35 38 44 47 47 50 53 56 59

Using the same mentality as above the only legal possibility would be 41 in 65 out, which is impossible

26 29 32 35 38 41 44 50 53 53 56 59
26 29 32 35 38 41 44 47 50 56 59 59

These last two are clearly impossible.

<u>29-62</u> implies an original jury of 29 32 35 38 41 44 47 50 53 56 59 62 with one duplicate over 40 and one removed.

The total is 546 so we need a drop of 30. Thus we need an extra 29 and remove 59 or an extra 32 and remove 62 both impossible.

<u>32-65</u> implies that the original jury was 32 35 38 41 etc. up to 65. They total 6 x 97 = 582 requiring a drop of 66 – too much.

Thus the original jury is 26 29 32 38 41 41 44 47 50 53 56 59
The replacement jury is 26 29 32 35 38 41 41 44 47 50 53 56

And the twelve ages are 26 29 32 35 38 41 44 47 50 53 56 59

Answer: 1, 2, 1, 2, 2 & 1 pesos

Each of the coins is of either of unit weight 1 (one peso) or unit weight 2 (two pesos). Starting from the top let the 6 coins be of weights w1 to w6 and their diameters 'd'. The coins are leaning in one direction such that they are on the verge of toppling.

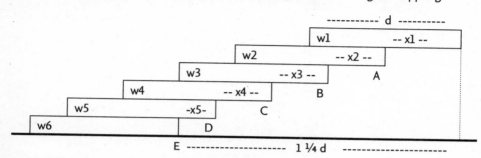

Considering point A then the centre of gravity of w1 must be at the outer edge of w2 so x1 = d/2

For point B the combined centre of gravity of w1 and w2 must be on the edge of w3 i.e. directly above point B. So taking the moments of force about point B gives –

x2w1 + (x2 – d/2)w2 = 0 : x2 = dw2/2(w1+w2)

Similarly for point C – (x2+x3)w1 + (x2+x3-d/2)w2 + (x3 -d/2)w3 = 0

Rearranging and substituting for x2 gives – x3 = dw3/2(w1+w2+w3)

And for points D & E : x4 = dw4/2(w1+w2+w3+w4), x5 = dw5/2(w1+w2+w3+w4+w5)

x1 + x2 + x3 + x4 + x5 = 1¼d

d{1 + w2/(w1+w2) + w3/(w1+w2+w3) + w4/(w1+w2+w3+w4) + w5/(w1+w2+w3+w4+w5)}/2 = 1¼d

w2/(w1+w2) + w3/(w1+w2+w3) + w4/(w1+w2+w3+w4) + w5/(w1+w2+w3+w4+w5) = 1.5

The most significant figure is w1 and a quick check shows that this must be of weight 1. (If w1 is 2 then for both extremes, w2-w5 =1 and w2-w5 =2, the total value is < 1.5, so no solutions if w2 is equal to 2)

With w1=1 and the four remaining coins w2-w5 comprising 2 x 1p and 2 x 2p then the largest fraction denominator is 7. With a numerator of 1 or 2 this can, with the other components, never total 1.5 so this combination can be excluded. Likewise for w2-w5 all being 1s giving the largest denominator 5. This leaves the following combinations –

w1	w2	w3	w4	w5		w1+w2	w1+..+w3	w1+..+w4	w1+..+w5	
1	2	1	1	1		3	4	5	6	(a)
1	1	2	1	1		2	4	5	6	(a)
1	1	1	2	1		2	3	5	6	(a)
1	1	1	1	2		2	3	4	6	(b)
1	1	2	2	2		2	4	6	8	(a)
1	2	1	2	2		3	4	6	8	(c)
1	2	2	1	2		3	5	6	8	(a)
1	2	2	2	2		3	5	7	9	(a)

(a) Can be eliminated on inspection

(b) $1/(1+1) + 1/(1+1+1) + 1/(1+1+1+1) + 2/(1+1+1+1+2) = 1.208$

(c) $2/(1+2) + 1/(1+2+1) + 2/(1+2+1+2) + 2/(1+2+1+2+2) = 1.5$ – Correct solution.

The only solution for the top five coins is 1, 2, 1, 2 and 2, which totals 8, so for the six coins to add to an odd number then w6 must equal 1.

From top to bottom the coins are 1, 2, 1, 2, 2 and 1.

Answer: 2, 6, 7, 9, 18 and 42

The reciprocals of the eleven given squad numbers sum to 4777/2520, so to bring the total up to 2 the three unknown reciprocals must sum to 263/2520. Call this value R. To get an answer with 2520 as the denominator, the Lowest Common Multiplier (LCM) of the unknowns must be 2520 = 5 x 7 x 8 x 9, or a multiple thereof. So each of those prime powers must divide at least one of the unknowns.

No unknown can be a multiple of three of those factors, because it would be too big. So for three unknowns to deal with the four factors, at least one unknown must be a product of two of the factors or a multiple thereof, and of course must be less than 100. So there are nine cases, as examined below; each of the nine possible reciprocals is subtracted from R in turn, and reduced to lowest terms by cancelling any common factors.

We need to find suitable integers a & b satisfying $1/a + 1/b = m/n$; equivalently, $n(a+b) = mab$. As it happens (see below) m is prime in every one of the nine cases; so it must divide $a+b$. Write $a+b = km$, and thereby also $ab = kn$. So $nkm = ma(km-a)$, giving the quadratic $a^2-kma+kn = 0$ (and the same equation would hold for b). 'Complete the square' to write this equivalently as $(2a-km)^2 = k^2m^2-4kn$.

For there to be any hope of an integer solution, the right hand side (call it D) must be non-negative, requiring $k \geq 4n/m^2$; it must also be a perfect square. Its square root can be taken as positive or negative, leading to candidate values of a & b as $\frac{1}{2}(km\pm\sqrt{D})$. Write $k = g^2h$ where any square factors of k are gathered into g. Then $D = g^2h(g^2hm^2-4n)$; h occurs explicitly outside the bracket, so for this to be a square, h must also divide the bracketed term, and since it certainly occurs in the first part it must also divide the second part, i.e. h must divide $4n$. Also, since $km = a+b \leq 197$ to avoid being too big, $k \leq 197/m$. Now each of the nine cases can be examined in turn, using various points from the foregoing.

R-1/35 = 191/2520. So a & b must be between 92 & 99, $k = 1$, but then ab cannot equal 5x7x8x9.

R-1/40 = 5/63. $k > 10$ and $k < 40$. Possible values of k are: $12=2^2\times3 \rightarrow D = 24^2 \rightarrow$ a & b = 42 & 18 which is a legal solution; $14 \rightarrow D = 14^2\times7$, not square; $16=4^2 \rightarrow$ $D = 8^2\times37$, not square; $18=3^2\times2 \rightarrow D = 18^2\times11$, not square; $21 \rightarrow D = 21^2\times13$, not square; $24=2^2\times6 \rightarrow D = 12^2\times58$, not square; $25=5^2 \rightarrow D = 5^2\times373$, not square; $27=3^2\times3 \rightarrow$ $D = 9^2\times141$, not square; $28=2^2\times7 \rightarrow D = 112^2 \rightarrow a$ & b = 126 & 14, too large; $32=4^2\times2 \rightarrow D = 8^2\times274$, not square; $36=6^2 \rightarrow D = 108^2\times2$, not square.

R-1/45 = 23/280. $k > 2$ and $k < 9$. Possible values for k are: $4=2^2 \rightarrow D = 4^2\times249$, not square; $5 \rightarrow D = 5^2\times305$, not square; $7 \rightarrow D = 7^2\times369$, not square; $8=2^2\times2 \rightarrow D = 8^2\times389$, not square.

R-1/70 = 227/2520. $k < 1$, which is impossible.

R-1/80 = 463/5040. $k < 1$, which is impossible.

R-1/90 = 47/504. $k \geq 1$ and $k \leq 4$. Possible values for k are: $1 \rightarrow D = 193$, not square; $2 \rightarrow D = 2^2\times1201$, not square; $3 \rightarrow D = 3^2\times1537$, not square; $4 \rightarrow D = 4^2\times1705$, not square.

R-1/56 = 109/1260. $k \geq 1$ and $k \leq 1$. $k = 1 \rightarrow D = 6841$, not square.

R-1/63 = 223/2520. $k < 1$, which is impossible.

R-1/72 = 19/210. $k \geq 3$ and $k \leq 10$. Possible values for k are: 3 → D = 27^2 → a & b = 15 & 42, repeats 15; 4 → D = 4^2x151, not square; 5 → D = 5^2x193, not square; 6 → D = 6^2x221, not square; 7 → D = 7^2x241, not square; 8 → D = 128^2 → a & b = 140 & 12, too big; 9=3^2 → D = 3^2x209, not square; 10 → D = 10^2x277, not square.

So the only possible set of squad numbers is the given values 2, 3, 4, 5, 6, 7, 8, 9, 15 with the three unknowns being 40, 42, 18.

If we add fractions in their lowest terms, and there is a prime power that divides the denominator of only one of them, then this prime power cannot cancel out in the result – because, when the terms are expressed with a denominator that is the LCM of all individual denominators, all but one of the numerators will have the prime power as a factor, so the sum of them cannot be divisible by the prime power.

All three multiples of 5 must be on the same side, because none of them can be on its own in a team. The reciprocals of these three sum to 7/24. A factor of the denominator is 8, and the only remaining reciprocal with this factor is 1/8, so 8 must be in the same team, and adding its reciprocal brings the total to 5/12. The denominator still has a factor of 4, and 4 itself is the only remaining number with this factor, so this is in the same team and the total goes to 2/3. That puts five numbers in the one team, with the total currently 1/3 short of 1, so 3 can be included in the same six-player team to make the total 1. This team therefore consists of 3,4,5,8,15,40.

As a check on the other team, (1/7 + 1/42) + (1/9 + 1/18) + 1/6 + 1/2 = 1/6 + 1/6 + 1/6 + 1/2 = 1. This team consisting of 2,6,7,9,18,42 is the answer that was asked for.

29 VALUED PLAYWRIGHTS

Answer: 67, 31 and 53

[In this solution a string of capitals represents the total of their values; e.g. PLAY=39]

P=4 L=8 A=3 Y=24

SHAW = SHAKESPEARE, so W = KESPEARE and is at most 26.
KESPEARE = EEE + 3 + 4 + KRS
We tabulate the possibilities:

	E	K/R/S	W	Available for I/N/D/O
(a)	1	2/5/6	23	7,9,10,11,...
(b)	1	2/5/9	26	6,7,10,11,...
(c)	1	2/6/7	25	5,9,10,11,...
(d)	2	1/5/6	25	7,9,10,11,...
(e)	2	1/5/7	26	6,9,10,11,...

Now
PIRANDELLO = PALL+ER+INDO= 4+3+8+8+ER+INDO,

and this has a minimum possible value of 60. Also SHAW has a maximum possible value of 9+25+3+26=63. Since their common value is prime it must be 61. Therefore ER+INDO=38 and this is only possible in
case (b) with E=1, R=2 and I/N/D/O = 6/7/10/12
or   case (c) with E=1, R=2 and I/N/D/O = 5/9/10/11.

But PINERO is prime so 7 + INDO − D is prime, and this only works in case (c) with
E=1, **R=2**, K/S=6/7, **W=25** and I/N/D/O=5/9/10/11(with D=5 or 11)

Also SHAW=61=S+H+3+25, so H=33 − S, making **S=7** and **H=26**. Also **K=6**

Now RATTIGAN = 61 = 8 + IN + TTG. But IN has minimum possible value of 14, so TTG has a maximum possible value of 39. Since only 12, 13, ... are available for G and T it follows that **T=12**, **G=15**, and I/N=5/9 (and **D=11**, making **O=10**).

For FRAYN=61 we need FN=32, so **N=9** and **F=23** (and **I=5**).

For BECKETT=61 we now need BC=29 so we must have B/C=13/16
But COWARD = C+51 is prime so **C=16** and **B=13**.

So
A=3, B=13, C=16, D=11, E=1, F=23, G=15, H=26, I=5, K=6,
L=8, N=9, O=10, P=4, R=2, S=7, T=12, W=25 and Y=24.

COWARD=16+10+25+3+2+11=67
PINERO=4+5+9+1+2+10=31
STOPPARD=7+12+10+4+4+3+2+11=53

Answer: 10

Suppose that the tray is arranged with its width equal to 15cm and its height equal to 16cm and that the rectangles are numbered 1 to 8, in ascending order of size. A 'Top or Bottom' set is a set of rectangles that can be placed along the top or bottom edge of the tray and a 'Left or Right' set is a set of rectangles that can be placed against the left or right edge of the tray, as shown in the diagrams below.

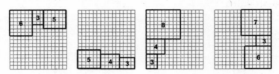

If rectangles 1 and 2 are not included, there are 18 different Top or Bottom sets and 22 different Left or Right sets:-

Top or Bottom	Left or Right
(8, 7)	(8, 7) : two sets
(8, 6) : two sets	(8, 6)
(8, 5)	(8, 5, 3)
(8, 4, 3)	(8, 4, 3) : three sets
(7, 6)	(7, 6, 3)
(7, 5, 3)	(7, 5, 4)
(7, 4, 3) : three sets	(7, 5, 3) : three sets
(6, 5, 4)	(7, 4, 3) : three sets
(6, 5, 3) : three sets	(6, 5, 4) : three sets
(6, 4, 3) : three sets	(6, 5, 3) : three sets
(5, 4, 3)	(6, 4, 3)

There are only seven ways of selecting two Top or Bottom sets and two Left or Right sets to fill all four edges of the tray:-

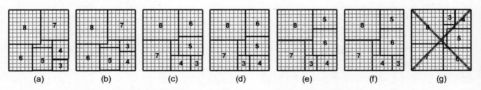

(a)　(b)　(c)　(d)　(e)　(f)　(g)

Configurations (a) to (f) have gaps in the middle into which rectangles 1 and 2 can be inserted, but configuration (g) does not. Rectangles 1 and 2 can be inserted into configurations (c) to (f) in two different ways, so there are a total of 10 solutions.

If rectangles 1 and 2 are included there are many more sets, but no solution is possible with any of these sets. Eight ways in which rectangles 1 and/or 2 can be included in a set are shown in diagrams (a) to (h) below but four are invalid because they leave a gap above rectangle 1 that cannot be filled. Configuration (b) is valid because 2 can be placed above 1; configuration (c) is valid because 2 is next to 3; configuration (d) is valid because 3 can be placed above 2; configuration (f) is valid because 1 is next to 2. Hence, if rectangle 1 belongs to a set it must be adjacent to 2 and if rectangle 2 belongs to a set it must be adjacent to 1 or 3, with sides of equal length touching.

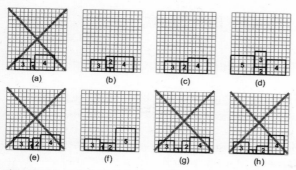

The diagrams below show that no solution is possible with 1 in a set and adjacent to 2. Inter-changing the positions of 1 and 2 in every diagram shows that no solution is possible with 2 in a set and adjacent to 1.

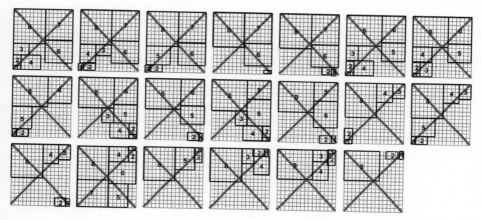

The diagrams below show that no solution is possible with 2 in a set and adjacent to 3.

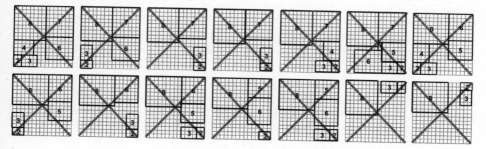

Answer: Rathripe, Transponder, Vexatious and Wergild

Because there are two actors that I dislike, and can stand no more than two films containing each of them, the greatest number of films that I could binge-watch today is four.

Consider pairs of films that I could watch in order to see Amerton twice; there are ten such pairs. For each pair I need to choose at most two other films to ensure I see each of my favourite actors (i.e. all except Ingleby and Joyford) at least twice.

Watching *Quaver* and *Rathripe* gives me no appearances by Blankney or Hendy, so I must see them both in two other films, which restricts me to two of *Underpassion*, *Wergild* and *Yarborough*; but none of those three gives me another film with Codrington in, to add to the one in *Rathripe* (but not *Quaver*), so this doesn't work.

Similarly, watching *Quaver* and *Statecraft* gives me no appearances by Blankney or Glanton, so I need to choose two of *Wergild*, *X-axis* and *Yarborough*; but none of those is a second Emstrey film.

Watching *Quaver* and *Vexatious* gives me no appearances by Codrington, Glanton or Hendy, and no films at all contain all three of those.

Watching *Quaver* and *Zeuxis* gives me no Blankney or Hendy, so I need to choose two of *Underpassion*, *Wergild* and *Yarborough*; but none of those three gives me the second film with Codrington in.

Rathripe and *Statecraft* don't have Blankney or Emstrey, so I must choose *Underpassion* and *Vexatious*, but they don't feature a second Glanton.

Rathripe and *Vexatious* don't have Hendy in, so I'd need to choose two of *Statecraft*, *Transponder*, *Underpassion*, *Wergild* and *Yarborough*. To get a second Codrington I'd need one of *Statecraft* and *Transponder*; for a second Emstrey, one of *Transponder* and *Underpassion*; for a second Fairwater, one of *Underpassion* and *Wergild*; and for a second Glanton, one of *Wergild* and *Yarborough*. The only way to achieve these conditions in only two films is to choose *Transponder* and *Wergild*. Then it can be seen that the four films *Rathripe*, *Transponder*, *Vexatious*, and *Wergild* together feature all the named actors exactly twice, satisfying my wish for at most two appearances by each of the actors I dislike and at least two by each of the other actors. And because these four films already provide two appearances by both of my disliked actors, I cannot view any more films than these.

Rathripe and *Zeuxis* don't feature Blankney, Dunino and Hendy, but only one film, *Yarborough*, features all three of them.

Statecraft and *Vexatious* don't feature Fairwater and Glanton, so I need two of *Rathripe*, *Wergild* and *X-axis*, but none of those features Emstrey a second time.

Statecraft and *Zeuxis* don't have Blankney and Fairwater, so I need two of *Underpassion*, *Wergild* and *X-axis*, but none of those has Dunino a second time.

Vexatious and *Zeuxis* leave me without Fairwater and Hendy, so I'd need to choose *Underpassion* and *Wergild*, but neither of them provides me with a second appearance of Codrington (and the four films taken together would also give me three appearances by my disliked Joyford).

So the solution stated above is the only collection of films that would meet my conditions.

32 THREE-CORNERED PROBLEM

Answer: 17

With N different numbers to choose from the number of triangles is:

X
X   X

N of these

X
X   Y

(X, Y different)

N(N – 1) of these

X
Y   X

(X, Y, Z different)

N(N – 1)(N – 2)/3 of these

So the number of different triangles is
T(N) = N + N(N – 1) + N(N – 1)(N – 2)/3
i.e. T(N) = N(N^2 + 2)/3

We can soon see that
T(odd) = odd & T(even) = even.

If my age is A then the given different facts about the odds and evens in {1, 2, ... , A} show that A must be odd, A = 2B + 1 say.

The number of triangles containing at least one odd = T(A) – no. containing no odd
                                                      = T(A) – no. containing all evens
                                                      = T(A) – T(B)

We are told this is odd and, as T(A) is odd, it follows that T(B) is even; i.e. B is even. So my age, 2B + 1, is one of 5, 9, 13, 17, 21, 25 or 29.

The number of triangles containing at least one multiple of four is

T(A) – T(no. in {1, 2, ... , A} not divisible by 4)

For this to be even we need an odd number of {1, 2, ... , A} to be not divisible by four; i.e. an even number of {1, 2, ..., A} must be divisible by four. So A must be one of 9, 17 or 25:

A	How many of 1 to A are not divisible by 4?	So, no of triangles containing a mult of 4 is	which equals	Divisible by 4?
9	7	T(9) - T(7)	3.83 – 7.17 = 130	No
17	13	T(17) - T(13)	17.97 – 13.57 = 908	**Yes**
25	19	T(25) - T(19)	25.209 – 19.121 = 2926	No

So only when A = 17 is the number of triangles containing a multiple of four equal to a multiple of four.

33   DIAL M FOR MARRIAGE

Answer: 7753222 and 7553332

There are 10!/6!4! ways of choosing four numbers from ten; that is 10.9.8.7/4.3.2 = 210. Once a group of four has been chosen, we shall call them a b c and d.

Then there are three arrangements for seven-digit numbers using those four:
 1) abcdddd   2) abccddd   3) abbccdd

Considering 1), there are four ways to choose which is the majority digit. They can be placed in any of the following positions:

1234 1235 1236 1237 1245 1246 1247 etc. 7!/4!3! = 35. Each can accommodate six different placings of the other three digits, so we are considering 4 x 35 x 6 = 840.

Considering 2) there are four ways to choose the majority digit and three for the one with only two representatives, twelve in all. Again, there are 35 arrangements for the majority digit, and 4!/2!2! = 6 for the runner up. The remaining two 'singles' can be placed either way so we are considering 12 x 35 x 6 x 2 = 5040.

Considering 3) there are four ways to choose the minority digit and it can be placed in any one of the seven positions. So if we look at the following table where there is a choice of picking two from six places for the second choice and two from four positions for the third:

a	b	c	d
7	6!/4!2!	4!/2!2!	
	=15	=6	

And there are 24 ways of choosing the order, we are considering 24 x 7 x 15 x 6 = 15120.

So the grand total is 840 + 5040 + 15120 = 21000 and it needs to be multiplied by 210 to give 4410000.
This factorises to 2^4 x 3^2 x 5^4 x 7^2 so 2 3 5 7 are the required numbers. The only perfect number in the range is 28 and this can only be achieved with 2 + 3 + 5 + 7 = 17 leaving 11 for repeats (2+2+7 or 3+3+5); the only solutions are 7753222 and 7553332 so these are the two numbers.

34 SNOOKERED

Answer: 26 feet

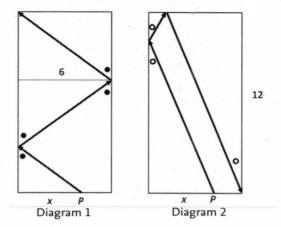

Diagram 1   Diagram 2

If we reflect the first leg of Diagram 1 in the left cushion, and the third leg in the right cushion we see that the path of the ball becomes the hypotenuse of a right-angled triangle with short sides $(x + 12)$ and 12 and hence with length $\sqrt{(x + 12)^2 + 12^2}$.

If we reflect the first leg of Diagram 2 in the left cushion, and the third leg in the top cushion we see that the path of the ball becomes the hypotenuse of a right-angled triangle with short sides 24 and $(x + 6)$ and hence with length $\sqrt{(x + 6)^2 + 24^2}$.

We want the second path to be 30% longer than the first, so
$$1.3 \times \sqrt{(x + 12)^2 + 12^2} = \sqrt{(x + 6)^2 + 24^2}$$

Let $(x + 12) = y$ and square both sides:
$$1.69\,(y^2 + 144) = (y - 6)^2 + 576$$

Multiply both sides by 100 and rearrange into a standard quadratic equation:
$$69y^2 + 1200y - 36864 = 0$$

This looks completely hopeless, but work out the discriminant:
$$\Delta = 1200^2 - 4 \times 69 \times -\,36864 = 11614464$$
And now $11616664 = 3408^2$

(which is amazing: the setter exploited 3-4-5 and 5-12-13 triangles and the 6 by 12 table to make the puzzle)
$$y = \frac{-1200 \pm 3408}{2 \times 69} = 16 \text{ or } -\frac{768}{23}$$
Now $y = (x + 12) = 16$ is the only answer that makes sense on the snooker table.

So $x = 4$, and the distance the ball moves on the second shot is $\sqrt{(4 + 6)^2 + 24^2}$ = 26 feet.

Answer: 5ft by 7ft

$\leftarrow$ r squares $\rightarrow$

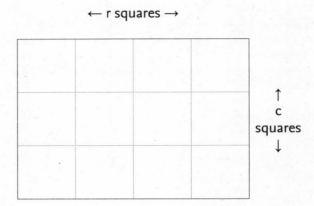

$\uparrow$
c
squares
$\downarrow$

Let the rectangular throw be r squares across and c squares down.

The total number of sewn joins between squares in the rectangular throw is 2cr − (c + r). This would also apply to any square throw when c = r.
For squares of one square foot the length of sewing is therefore 2cr − (c + r) ft. and the knitted area is cr ft.²

If the time to knit each square foot is 'k' and the time to sew each foot is 's' then the total time to complete the rectangular throw is s{2cr − (c + r)} + kcr

The dimensions of the square throw are r x r, so the total time to complete it is, substituting r for c in the above equation = s{2r² −2r} + kr².

The total time to complete each square foot of the rectangular throw is
[s{2cr − (c + r)} + kcr] / cr and that of the square throw [s{2r² −2r} + kr²] / r²

So [s{2r² −2r} + kr²] / r² = 1.02 [s{2cr − (c + r)} + kcr] / cr

Substituting k = 1.2s in the equation and simplifying, this becomes

c = 255r / (16r + 245) or
r = 980c / (1020 − 64c), so for r to be positive then c < 16

Checking values of c < 16, the only c/r solutions are 5/7 and 15/245.

As the number of squares used for c x r is less than 100 the solution is 5ft by 7ft.

Answer: 109

The lowest common multiple has at least 2 digits and the highest common factor has no more than 2 digits so they both equal a two-figure number x. Three different multiples of x cannot be less than x, 2x and 3x, respectively, so 3x is no greater than 99 and x is no greater than 33. Three different factors of x cannot be less than x/3, x/2 and x/1, respectively, so x/3 is at least 10 and x is at least 30. The three factors of x are in the range 10-30 and must be equal to 10, 15 and 30 so that x is 30. The three multiples of 30 are 30, 60 and 90.

Hence, there were 30 guests for dinner with ticket numbers 1, 2, ..., 30 and there were 6 possible seating arrangements:-
(1) 2 tables each with 15 guests
(2) 3 tables each with 10 guests
(3) 5 tables each with 6 guests
(4) 6 tables each with 5 guests
(5) 10 tables each with 3 guests
(6) 15 tables each with 2 guests

The sum of the 30 ticket numbers is 30(30 + 1)/2 = 465 and each prime sum was less than 150 so there were at least 4 tables, which rules out (1) and (2). Couples were seated with friends so there were at least 3 guests per table, which rules out (6). The prime sums were odd numbers (because all primes greater than 2 are odd) and 465 is also odd, so there was an odd number of tables, which rules out (4) and (5).

Hence, there were 5 tables each with 6 guests.

The lowest possible prime sum of 6 ticket numbers is 1 + 2 + 3 + 4 + 5 + 8 = 23 and the highest possible prime sum is 23 + 26 + 27 + 28 + 29 + 30 = 163. If the two prime sums are denoted p and q then either 3p + 2q = 465 or 4p + q = 465. If each prime number p in the range 23 to 163 is multiplied by 3, subtracted from 465 and divided by 2 to give q, there are no results for which q is a prime number. If each prime number p is multiplied by 4 and subtracted from 465 to give q, there are six results for which q is a prime number:-

p	q
79	149
89	109
101	61
103	53
107	37
109	29

If told that the larger prime is 101, 103, 107 or 149 the smaller prime can be determined as 61, 53, 37 or 79, respectively, but if told that it is 109 the smaller prime could be 89 or 29.

Hence, the larger of the two prime numbers is 109.

Answer: 397cm

Three edges meet at each of the eight corners so the edges alone cannot be completely traversed by staying on the edges. The bees must also travel on some face or space diagonals (still from 'corner to corner'). The number of path elements entering all but two corners (the start and finish corners) must be even, so we need at least three diagonal path elements. Since we are looking for a minimal total, we need only consider exactly three diagonal path elements.

The diagonal elements could be face or space diagonals. Since all space diagonals meet at the centre there cannot be two space diagonals in a path, because the paths only intersect at corners of the boxes.

The two possible topologies leading to minimal path-lengths are shown in the diagrams. The diagonal elements are either two opposite face-diagonals and one space diagonal or two opposite face-diagonals and another face-diagonal.

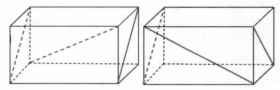

Note that if we use non-opposite face-diagonals the path will be longer than using both shortest opposite face-diagonals.

Because the total path length is an integer the diagonal elements must also be integer length so now we are looking for cuboids with distinct integer sides (because 'no face is square') with two non-equal integer diagonals at least one of which is a face-diagonal. Let the cuboid have dimensions $x \times y \times z$ such that $f = \sqrt{x^2 + y^2}$ is an integer face-diagonal.

The relevant Pythagorean triples are $(x, y, f) = (3,4,5)$, $(6,8,10)$, $(9,12,15)$, $(12,16,20)$, $(5,12,13)$, $(8,15,17)$.

The triple $(7, 24, 25)$ gives a box with sides $7 \times 24 \times z$, but since the volume is to be less than 1000, $z = 1,2,3,4,5$ and none of these values gives a second integer face/space-diagonal. All other primitive triples require $z \leq 2$ which cannot give an integer diagonal. Similar arguments dispose of $(15,20,25)$, $(10,24,26)$ and $(16,30,34)$.

(A) We look first for cuboids with integer space diagonals.
In a $3 \times 4 \times 12$ cuboid, the space diagonal length is $s = \sqrt{5^2 + z^2}$ so
$s^2 - z^2 = 25$ $(s–z)$ $(s +z)$ and we must have $(s - z) = 1$, $(s + z) = 25$ and $s = 13$, $z = 12$.
The cuboid is $3 \times 4 \times 12$ with volume 144, and we can't scale by 2 because that gives volume $1152 > 1000$.
In a $6 \times 8 \times z$ cuboid, the space diagonal length is $s = \sqrt{10^2 + z^2}$ so
$s^2 - z^2 = 100 = (s - z)(s + z)$ and we must have $(s - z) = 2$, $(s + z) = 50$ and $s = 26$, $z = 24$ with volume 1152.

In a $9 \times 12 \times z$ cuboid, the space diagonal length is $s = \sqrt{15^2 + z^2}$ so
$s^2 - z^2 = 225 = (s - z)(s + z)$ and we must have $(s - z) = 9$, $(s + z) = 25$ and $s = 17$, $z = 8$
with volume 864.

In a $12 \times 16 \times z$ cuboid, the space diagonal length is $s = \sqrt{20^2 + z^2}$ so
$s^2 - z^2 = 400 = (s - z)(s + z)$ and we must have $(s - z) = 2, 4, 8$, $(s + z) = 200, 100, 50$
and $(s, z) = (101,99), (52,48), (29,21)$ with volumes 19008, 9212, 4032.

In a $5 \times 12 \times z$ cuboid, the space diagonal length is $s = \sqrt{13^2 + z^2}$ so
$s^2 - z^2 = 169 = (s - z)(s + z)$ and $(s - z) = 1$, $(s + z) = 169$, $s = 85$, $z = 84$
and volume is 5040.

In an $8 \times 15 \times z$ cuboid $z = 144$, $s = 145$ with volume 17280.

The only cuboids with volume less than 1000 with an integer face-diagonal and
integer space-diagonal are $3 \times 4 \times 12$ and $9 \times 12 \times 8$.

(B) Now we look for cuboids with a second integer face-diagonal, by scaling up primitive
triples so that a short side can be part of a different triple.

x	y	z	volume	diagonals	
6	8	15	720	10, 17	(3,4,5) scaled up by 2 to meet (8,15,17)
9	12	5	540	15, 13	(3,4,5) scaled up by 3 to meet (5,12,13)
12	16	5	960	20, 13	(3,4,5) scaled up by 4 to meet (5,12,13)
36	15	8	4320 > 1000	39, 17	(5,12,13) scaled up by 3 to meet (8,15,17)
15	20	8	2400 > 1000	25, 17	(3,4,5) scaled up by 5 to meet (8,15,17)

So there are only five possible cuboids with volume less than one litre. The shortest
path lengths are:

x	y	z	minimum path length
6	8	15	$(6 + 8 + 15) \times 4 + 10 + 10 + 17 = 153$
9	12	5	$(9 + 12 + 5) \times 4 + 13 + 13 + 15 = 145$
12	16	5	$(12 + 16 + 5) \times 4 + 13 + 13 + 20 = 178$
3	4	12	$(3 + 4 + 12) \times 4 + 5 + 5 + 13 = 99$
9	12	8	$(9 + 12 + 8) \times 4 + 15 + 15 + 17 = 163$

So the required minimal integer path length is $153 + 145 + 99 = 397$.

Answer: 5,6,9,6,5,6

Write the lengths of the three Arts series as a_1, a_2, a_3, the two Biography series as b_1, b_2, and the one Comedy series as c_1. The possible schedules can be represented by the arrangements of those symbols where the a_i and the b_i are in order, with the same labels (a, b, c) never being adjacent, and with the same label not appearing at both ends (because then all of that topic's series could not be moved forwards or backwards in time). This gives possibilities as $a_1 b_1 a_2 b_2 a_3 c_1$, $a_1 b_1 a_2 c_1 a_3 b_2$, $a_1 c_1 a_2 b_1 a_3 b_2$, $b_1 a_1 b_2 a_2 c_1 a_3$, $b_1 a_1 c_1 a_2 b_2 a_3$, $c_1 a_1 b_1 a_2 b_2 a_3$.

Now the schedules have to be paired, to represent the shuffling that took place. No pairing can have both schedules from the first three, because then $a1$ would not move; and similarly for the last three, because of a_3. That leaves nine potential pairings, which consist of one from the first three and one from the last three. Note that reversing the order of the two schedules in a pairing is simply equivalent to multiplying all the values of the shifts by -1, or, equivalently, multiplying the equations for the series lengths by -1. So for simplicity at this stage it can be taken that the schedule from the first three possibilities above is the 'before' situation and the schedule from the last three is the 'after'.

Observe that if a pairing has two or more adjacent symbols occurring in the same order in the 'before' and 'after' schedules, then those symbols represent differently-labelled series that have shifted by the same amount, which contradicts the conditions. This rules out five of the pairings: $a_1 b_1 \mathbf{a_2 b_2} a_3 c_1$ & $b_1 a_1 c_1 \mathbf{a_2 b_2} a_3$, $a_1 b_1 a_2 b_2 a_3 c_1$ & $c_1 a_1 b_1 a_2 b_2 a_3$, $a_1 b_1 \mathbf{a_2 c_1 a_3} b_2$ & $b_1 a_1 b_2 \mathbf{a_2 c_1 a_3}$, $a_1 b_1 \mathbf{a_2 c_1 a_3} b_2$ & $c_1 a_1 b_1 \mathbf{a_2 b_2 a_3}$, and $\mathbf{a_1 c_1} a_2 b_1 a_3 b_2$ & $b_1 \mathbf{a_1 c_1} a_2 b_2 a_3$ (the offending neighbours being identified in **bold**).

Now write A, B & C for the amount by which the Arts, Biography and Comedy series are being shifted. Then equations can be set up in turn for the remaining four pairings, relating the size of the shift of each series to the sizes of the other series that it must 'jump over' backwards or forwards. Obviously for a legal solution all the series lengths must be positive. The teaser conditions give the three shifts as A = -6, B = 20, C = -21.

For $a_1 b_1 a_2 b_2 a_3 c_1$ going to $b_1 a_1 b_2 a_2 c_1 a_3$, the simultaneous equations that have to be solved are $A = b_1$, $B = -a_1$, $A = b_2$, $B = -a_2$, $A = c_1$, $C = -a_3$, and the solution is straightforwardly $a_1 = a_2 = -B$, $a_3 = -C$, $b_1 = b_2 = A$, $c_1 = A$. But this doesn't work because it would for instance make $b_1 = A = -6$, and reversing the order of the pairing doesn't work because then $a_3 = (-1)\times(-C) = -21$.

For $a_1 b_1 a_2 c_1 a_3 b_2$ going to $b_1 a_1 c_1 a_2 b_2 a_3$, the simultaneous equations that have to be solved are $A = b_1$, $B = -a_1$, $A = c_1$, $C = -a_2$, $A = b_2$, $B = -a_3$, and the solution is straightforwardly $a_1 = a_3 = -B$, $a_2 = -C$, $b_1 = b_2 = A$, $c_1 = A$. But this doesn't work because it would for instance make $a_1 = -B = -20$, and reversing the order of the pairing doesn't work because then $a_2 = (-1)\times(-C) = -21$.

For $a_1c_1a_2b_1a_3b_2$ going to $b_1a_1b_2a_2c_1a_3$, the simultaneous equations that have to be solved are $A = b_1$, $C = a_2+b_1+b_2$, $A = b_1+b_2-c_1$, $B = -a_1-a_2-c_1$, $A = b_2$, $B = -a_2-a_3-c_1$, and the solution is $a_1 = a_3 = A-B-C$, $a_2 = C-2A$, $b_1 = b_2 = A$, $c_1 = A$. This doesn't work because it would for instance make $c_1 = A = -6$; but reversing the order of the pairing does work, with $a_1 = a_3 = (-1)\text{x}(A-B-C) = 6+20-21 = 5$, $a_2 = (-1)\text{x}(C-2A) = 21-12 = 9$, $b_1 = b_2 = c_1 = (-1)\text{x}A = 6$. So the lengths of the series after the shuffle are $a_1c_1a_2b_1a_3b_2$, which gives numbers for a solution.

For $a_1c_1a_2b_1a_3b_2$ going to $c_1a_1b_1a_2b_2a_3$, the simultaneous equations that have to be solved are $A = c_1$, $C = -a_1$, $A = b_1$, $B = -a_2$, $A = b_2$, $B = -a_3$, and the solution is straightforwardly $a_1 = -C$, $a_2 = a_3 = -B$, $b_1 = b_2 = A$, $c_1 = A$. But this doesn't work because it would for instance make $a_2 = -B = -20$, and reversing the order of the pairing doesn't work because then $a_1 = (-1)\text{x}(-C) = -21$.

So there is a unique solution of 5,6,9,6,5,6 as the eventual series lengths. The Arts series were originally scheduled for weeks 7-11, 18-26 & 33-37 but moved to 1-5, 12-20 & 21-31; the Biography for 1-6 & 12-17 but moved to 21-26 & 32-37; and the Comedy series for 27-32 but moved to 6-11.

39 DIGITAL DAISY-CHAINS

Answer: 583

The large number

The first letter of each digit is the last letter of another, and vice-versa.
Therefore SIX and SEVEN are ruled out (no digit ends S).
So NINE (and NOUGHT and NONE) are now ruled out (no remaining digit ends N).
FOUR, FIVE and ZERO are ruled out by their first letter.
That leaves ONE, TWO, THREE and EIGHT.
We can have 18..., 21..., 38..., 82... or 83....

If there is no 1 in the number then we must have 3838...38 (or 8383...83). In both cases the sum of the digits is a multiple of 11.

If there is no 3 in the number we must have 182182...182 (or 821821...821 etc), where again the sum of the digits is a multiple of 11.

To introduce a 3 it must be sandwiched between two 8s; i.e. take any of the previous possibilities and replace any 8 by 838 any number of times. This adds 11 to the digit total each time.

Hence my large number has a digit sum divisible by 11.

The digit sum

Since the digit sum is not a looped daisy-chain we can now use any non-zero digit except FOUR or SIX to start this daisy-chain sum. Clearly no two-figure multiple of 11 is a daisy-chain so we must have one of

18-, 21-, 38-, 58-, 70-, 79-, 82-, 83-, 98-, 182-, 183-, 218-, 382- or 383-

(The next highest would be 582- which is more than 520 lots of 11s and the original number would have more than 1000 digits.)

To be multiples of 11 these would have to be

187, 385, **583**, 704, 792, 825, 836, 1826, 1837, 2189, 3828 or 3839

The only one of these which is a daisy-chain is **583**.

[e.g. 182182182182.......182 with 53 repeats of 182.]

40   HALL OF RESIDENCE

Answer: Spanish, Gloucester

A table helps

1	2	3	4	5
		Chris		
	Spanish			
Not Glos				Colchester

And these can be filled in at once from the facts. Then it is a matter of trying to fit everything else in.

The only solution is

1	2	3	4	5
Dave	Andy	Chris	Bill	Ed
English	Spanish	German	French	Italian
Brighton	Gloucester	Reigate	Dunstable	Colchester

So, Andy from Gloucester is the student of Spanish.

41 ONE OF A KIND

Answer: 49, 53, 216, 780

It is worthwhile listing possible cubes and squares together with small triangular numbers:

cubes	8	27	125	216	512				
squares	4	9	16	25	49	81	169	196	256
	289	324	361	529	576	625	784	841	961
triangular no	3	6	10	15	21	28	45	78	91

There must either be one single-digit number or two 2-digit numbers as we have ten digits.

Considering the numbers from consecutive strips, we have three possibilities:

(a)	AC	BD	xxx	xxx	A=1 to 8, B=A+1, C=6 to 9, D=1 to 4
(b)	A9C	B0D	xx	xx	A=1 to 7, B=A+1, C=6 to 8, D=1 to 4
(c)	A9C	B0D	xxx	x	ditto

The cube cannot be 8 and it is easy to show it cannot be 27 (remembering that the smallest winning number isn't a cube):

13	27	x9x	x0x	No square possible from remaining digits.
15	27	x9x	x0x	ditto

It follows that we must have a 3-digit cube and it must contain the digits 1 and 2.

(a)	37	45	216	xxx	No square possible from remaining digits.
	49	53	216	xxx	A triangular number is required from digits 8,7,0.

Early cases of triangular numbers can be used to find the general formula $N = n(n + 1)/2$. This readily shows that the triangular numbers in the 700-900 region are 703, 741, **780**, 820

Therefore 49, 53, 216, 780 is a possible solution.

(b)	No cube is possible for x9x or x0x.				
(c)	4	216	A9C	B0D	No suitable value for C.
	4	125*	796	803	803 is not prime (and 796 is not a triangular number).
	(*or 512)				
	prime cube	A9C	B0D	No square possible.	

42 CONNECT FOUR

Answer: 13, 33, 37 and 63

Let the four numbers be a, b, c & d
If the reciprocal of a is the repetend of b*c*d then a times (b*c*d) is a number composed of all 9s.

> Example if a = 27 : 1 / a = 0.037037.... (ie repetend of 037) then
> 10^3 / a = 37.037037.......
> 10^3 / a - 1 / a = 37 ; which in the solution = (b*c*d)
> So   a * (b*c*d) = 10^3 − 1 = 999
> Similarly for other repetends, with different lengths requiring different numbers of 9s

As each two digit number has at least one of the digits a 3 then the product of all of these four numbers must be between about 35500000 (93*83*73*63) and about 278000 (13*23*30*31). The number of 9s is therefore either 6 or 7 ie 999,999 or 9,999,999.

Factoring 999,999 into its prime factors gives 3^3, 7, 11, 13 and 37.
Factoring 9,999,999 gives 3^2 but then no other prime factor below 12 exists so no two-digit number can be made by multiplying by 9. Likewise no prime factors below 34 exist and as two such factors are required to form two two-digit numbers when multiplied by the two 3s there can be no solutions for 9,999,999.

Using the prime factors of 999,999 (ie 3, 3, 3, 7, 11, 13, 37) the following table shows all of the four two-digit factors which can be created and which multiply to 999,999.

Prime Factors of 999,999	Combinations of Four Two Digits Numbers					
3^3	11	11	13	13	**13**	21
7	27	37	21	27	**33**	33
11	37	39	37	37	**37**	37
13	91	63	99	77	**63**	39
37						

Only the set of 13, 33, 37 & 63 has at least one 3 in each number, with the following table showing that these also satisfy the other requirements.

Two digit number	13	33	37	63
Reciprocal of two digit number	0.076923076923..	0.030303030303..	0.027027027027..	0.015873015873..
Remaining two digit no's	33, 37 & 63	13, 37 & 63	13, 33 & 63	13, 33 & 37
Product of remaining no's	76,923	30,303	27,027	15,873

43   IN THE SWIM

Answer: 13/75, 7/15, 3/5, 19/25

They can't have swum 1 length as the single fraction would not sum to an integer. They can't have swum 2 lengths as the two denominators would be the same. If they swam 3 lengths the fractions would lie in the intervals (0,1/3), (1/3,2/3), (2/3,1) and the total would lie in the interval (1,2) and could not be an integer. We seek 4 length solutions and the intervals argument shows that the integer total of the fractions is 2.

The odd two-digit common denominator d has to have 4 factors (one for each length) not including 1, so at least 5 factors. The possibilities are $d = 3^4$, $3^2.5$, 3.5^2, $3^2.7$, $3^2.11$

(A) $d = 3^4$: one denominator is 3, with fraction 2/3. The other (different) prime numerators are odd. When the fractions are given the odd common denominator, 2/3 will have a new even numerator and the other three new numerators will be odd so the total of the new numerators will be odd. But the total of the fractions is 2, so the total of the new numerators has to be even: contradiction. This '2/3 argument' will be used repeatedly: 2 cannot be a numerator.

(B) $d = 3^2.5$: the potential denominators are 3, 5, 9, 15, 45, but the '2/3' argument removes 3 and no numerator can be 2. So one fraction is 3/5 (0.6). Then 5/9 (0.56) is in the same length as 3/5 so another fraction must be 7/9 (0.77). 11/15 (0.73) and 13/15 (0.87) are in the same lengths as 3/5 and 7/9 and there is no solution.

(C) $d = 3.5^2$: the potential denominators are 3, 5, 15, 25, 75, but the '2/3' argument removes 3 and no numerator can be 2. So one fraction is 3/5 (0.6) . 7/15 (0.47) or 13/15 (0.87) are then possible. [11/15 (0.73) in 3rd length with 3/5]
3/5, 7/15 allows 19/25 (0.76) or 23/25 (0.92) and summing to 2 gives
(3/5, 7/15, 19/25, 13/75) or (3/5, 7/15, 23/25, 1/75)
3/5, 13/15 allows 7/25 (0.28) or 11/25 (0.44) and summing to 2 gives
(3/5, 13/15, 7/25, 19/75) [19/75 = 0.76 in 4th length with 13/15 = 0.87] or
(3/5, 13/15, 11/25, 7/75)

(D) $d = 3^2.7$: the potential denominators are 3, 7, 9, 21, 63 but the '2/3' argument removes 3 and no numerator can be 2. 3/7 (0.43), 5/7 (0.71), 5/9 (0.55), 7/9 (0.77), 5/21 (0.24), 11/21 (0.52), 13/21 (0.62), 17/21 (0.81), 19/21 (0.90). Because the set of numerators and denominators have no repeat, we see there are only two combinations that might work: (3/7, 5/9, 17/21, 13/63), but not (3/7, 5/9, 19/21, 7/63) [7/63 = 1/9 is not 'in the lowest terms']

(E) $d = 3^2.11$: the potential denominators are 3, 9, 11, 33, 99 but the '2/3' argument removes 3 and no numerator can be 2. 5/9 = 0.55, 7/9 = 0.77, 3/11 = 0.28, 5/11 = 0.46, 7/11 = 0.63

5/9, 3/11 with a] 7/33 = 0.21 b] 29/33 = 0.88 c] 31/33 and summing to 2 gives a] 95/99 or b] 29/99 (repeated 29) or c] 23/99 giving a solution
23/99 = 0.23, 3/11 = 0.28, 5/9 = 0.55, 31/33 = 0.93

7/9, 3/11 with a] 5/33 = 0.16, b] 17/33 = 0.51 c] 19/33 = 0.58, d] 23/33 = 0.70 leaves a] 79/99 (with 7/9 in 4th length) or b] 43/99 (with 3/11 in 2nd length) c] 37/99 (with 3/11 in 2nd length) d] $\underline{25}$/99.

7/9, 5/11 with a] 17/33 = 0.51, b] 19/33 = 0.58, c] 23/33 = 0.70 leaves a] $\underline{25}$/99, b] $\underline{19}$/99 (same numerator as 19/33), c] $\underline{7}$/99 (same numerator as 7/9)

We have found four solutions:
(13/63, 3/7, 5/9, 17/21), (13/75, 7/15, 3/5, 19/25), (7/75, 11/25, 3/5, 13/15), (23/99, 3/11, 5/9, 31/33)
(0.21, 0.43, 0.56, 0.81), (0.17, 0.47, 0.6, 0.76), (0.09, 0.44, 0.6, 0.87), (0.23, 0.28, 0.55, 0.93)
Tessa's fractions contain the one that is closest to an end 19/25 = 0.76, just after the third turn.

Tessa's fractions are 13/75, 7/15, 3/5, 19/25.
[We don't need to consider 5 lengths for the same reason as 3 lengths. 6 or more lengths requires an odd common denominator with at least 7 factors and this is at least 3.5.7 = 105.]

VILLAGE SIGNPOSTS

Answer: Carlton, Grafton, Barton

Label the signposts 1 to 7 as they appear in the conditions. For village names use the initial letters, or the signpost number until the name is deduced; if neither of these is known yet, use '#' in the diagrams. Overall maps could be drawn equivalently with any region as the outermost one, but with the interconnections being identical. Here the maps are labelled as i, ii, etc. and correspond to the paragraphs below.

(i) C & F both adjoin 3 & 6, so part of the map is as shown in the diagram. There are 6 villages in the map now, leaving just one more to be placed. Now look at the different cases of how the 'loose ends' could connect.

(ii) If 6 joins E then E, C & 3 must all join the seventh village (notated '#'), forcing B to join both E & F. But then B's neighbour order of E F … doesn't match any of the given conditions.

(iii) If 6 joins 3 then C must join F, and so the seventh village must join E F B in clockwise order, and that doesn't match any of the conditions.

(iv) If 3 joins B then C, B & 6 (and no fourth village) must all join the seventh village in that clockwise order. Now, E joins only 3 so far, and so must join at least two others, which now can only be F & B. So B has signpost 5 because it has four neighbours and signposts 3 & 6 have already been used. And since E is a known neighbour, D & G from this signpost can be placed. Then all the village names correspond to signposts as A=6, B=5, C=7, D=1, E=2, F=4 and G=3, so this is a valid arrangement (see diagrams (ix) & (x)).

(v) If the loose ends from 6 & 3 join none of the villages placed so far, they must all join the seventh village, and then C must also join it. The three as yet unplaced village names are A, G & D and they must all adjoin C, and in that clockwise order because that is the only order that matches a signpost (number 7). This gives three sub-cases.

(vi) If the situation is that 6=A, 3=G and the seventh village is D, then D's neighbours are A, C & G and perhaps a fourth, in that clockwise order, which fits only with signpost 4. This only gives three neighbours, and so each of B, F & E must join the other two. But then the clockwise neighbour order of B doesn't match any of the signposts.

(vii) If the situation is that 6=G, 3=D and the seventh village is A, then A's neighbours are G, C & D and perhaps a fourth, in that clockwise order, which doesn't match with any of the signposts.

(viii) If the situation is that 6=D, 3=A and the seventh village is G, then G's neighbours are D, C & A and perhaps a fourth, in that clockwise order, which doesn't match with any of the signposts.

(ix) & (x) So there is a unique arrangement fitting the conditions, as shown. Walking along roads using the 'rightmost' method results in going round the 'faces' of the map (which includes the area outside the map). Starting at D, and depending on which of its roads is begun with, would produce DBAD, DACD or DCGBD. Of these, DCGBD is the longest, and no other rightmost-only circuit anywhere is longer (i.e. no other 'face' has more than four edges); so this gives the answer to the question as posed in the teaser.

45 FACE VALUE

Answer: 1188

Plato's regular solid, having eight vertices, must be a cube. Divide the cube's six faces into three opposite pairs with numbers (a, a'), (b, b') and (c, c') on them.

Consider the expression P = (a + a') × (b + b') × (c + c'), the product of the sums of the pairs of numbers on opposite faces.

Expanding these three brackets by the distributive law yields the sum of 8 terms, each of which is the product of an "a" with a "b" and a "c". Because we paired opposite faces, these 8 products are just the labels on the vertices of the cube and so their sum is equal to the product P. Since the a's, b's and c's are distinct digits, their largest combined values are 4,5,6,7,8,9, and it follows that the new sum (the value of P after Plato's rearrangement) is at most $(9 \times 9 \times 8 \times 8 \times 7 \times 7)/(3 \times 3 \times 3) < 9^5 = 3^{10}$.

Consequently the new sum must be 1024, the 10th-power of 2. Each of the brackets in P is therefore a power of 2 between 4 and 16. The only possibilities are 4 × 16 × 16 and 8 × 8 × 16. Since 7 + 9 is the only pairing for 16 and the numbers a, a', b, b', c, c' are all different, the first possibility is ruled out. The only viable solution for the 8 × 8 × 16 option is P = (2 + 6) × (3 + 5) × (7 + 9) and, as Plato said, {2,3,5,6,7,9} is the *only* solution for the six numbers on the faces when the new sum is 1024; these are presumably the ones Eudoxus said he had discovered.

In addition to the solution shown above, there are fourteen further arrangements of Plato's six numbers on the faces, and these give the following values for the original sum:

(2 + 3) × (5 + 6) × (7 + 9) = 880
(2 + 3) × (5 + 7) × (6 + 9) = 900
(2 + 3) × (5 + 9) × (6 + 7) = 910

(2 + 5) × (3 + 6) × (7 + 9) = 1008
(2 + 5) × (3 + 7) × (6 + 9) = 1050
(2 + 5) × (3 + 9) × (6 + 7) = 1092

(2 + 6) × (3 + 5) × (7 + 9) = 1024 (unique solution)
(2 + 6) × (3 + 7) × (5 + 9) = 1120
(2 + 6) × (3 + 9) × (5 + 7) = 1152

(2 + 7) × (3 + 5) × (6 + 9) = 1080
(2 + 7) × (3 + 6) × (5 + 9) = 1134
(2 + 7) × (3 + 9) × (5 + 6) = 1188*

(2 + 9) × (3 + 5) × (6 + 7) = 1144
(2 + 9) × (3 + 6) × (5 + 7) = 1188*
(2 + 9) × (3 + 7) × (5 + 6) = 1210

The 13 values of the original sum not equal to 1024 admit at least four, and in the case of 900, sixteen distinct solutions for the numbers on the faces.

If Plato tells Eudoxus the original sum, he can work out the original number pairings except in the starred cases, where there are two possible arrangements of the face numbers, so the original sum must be 1188.

46 LUCKY PROGRESSION

Answer: 306, 918 and 2754

Let the first written number be N and my lucky number be L. Then the three written numbers use ten digits and are N (divisible by L), LN and L^2N

L cannot end in a 0, 1, 5 or 9 because they make the units digits of the three numbers contain a repeat. Similarly, if L ends in a 4 or 6 then N is even and the three numbers will have a repeat in their units digits. Also N cannot end in a 0 or 5.

Case (a): L>10

If L>10 then N must have at least 2 digits, so to give ten digits overall N, LN and L^2N must have 2, 3 and 5 digits respectively.

L	N (to give L^2N 5 digits)	N (to give LN 3 digits)	N/LN/L^2N
12	72, 84 or 96	72	72/864/10368 x
13	78 or 91	-	
17	51 or 68	51	51/867/14739 x
18	36, 54 or 72	36 or 54	36/648/... x or 54/972/17496 x
>22		23, 27 or 28	23/529/...x or 27/729/...x or 28/784/...x

Case (b): L<10

The numbers N, LN and L^2N must have 3, 3 and 4 digits respectively.

Since the three numbers use digits adding to 45 (a multiple of 9) it follows that
$(1 + L + L^2)N$ = sum of the three numbers = multiple of 9

L	$X=(1 + L + L^2)N$ (divisible by 9)	N (For 3-dig N, LN and 4-dig L^2N)
8	X=73N, so N divisible by 9 (and so by 72)	-
7	X=57N, so N divisible by 3 (and so by 21)	105 or 126, but both give a digit re-occurring in 7N
3	X=11N, so N divisible by 9	126 153 162 189 198 207 216 234 243 261 279 297 306 324.
2	X=7N, so N divisible by 9 (and so by 18)	306 324 342 378 396 432 468 486

But for L=3 each possibility except 169, 261, and 306 leads to the numbers N and LN having a digit in common. Of these remaining three only 306 works giving
306 918 2754

For L=2 in each case except 486 the numbers N and 2N include a repeated digit, and for N=486 we have $L^2N=1944$. So no N works in this case.

47 FOUR-SIDED DICE GAME

Answer: 2, 13 & 17

With one throw of both dice totals of 2 to 8 are possible. There are 16 throw combinations with some combinations giving the same totals. The frequency of each of these totals is shown in table 1. Likewise for two throws of both dice totals of 4 to 16 are possible with 256 combinations. Using the one-throw frequencies, those for the two throw totals can easily be calculated, shown in table 2 below. Because of their symmetry (as in table 1) only frequencies for the first 7 totals (of 13) need to be calculated.

The probability of Nia winning is the multiplication of a) and b) below
a) The number of possibilities of her second and third throw totals, when added to her first throw total, exceeding or equaling the target (let this be N(act)), divided by the number of total combinations of second and third throws = N(act) / 256
b) The number of possibilities of Rhys <u>not</u> achieving the above after his last throw = 1 - R(act) / 16

So the probability of Nia winning = (N(act) / 256) × (1 - R(act) / 16)
After the score adjustments (let these relevant numbers be N(adj) and R(adj)) Nia has a 35 times greater chance of winning. As both have a chance of winning, adjusting both their scores must result in both a) and b) above changing. The factors of 35 (excluding 1 which won't change probabilities), are 5 & 7 (but see note 1 below re 7/3 × 15/1).
So 35 × {(N(act) / 256) × (1 − R(act) / 16)} = (N(adj) / 256) × (1 − R(adj) / 16)
35{N(act) × (16 − R(act))} = N(adj) × (16 − R(adj))
Either: N(adj)/N(act) = 7 & (16−R(adj))/(16−R(act)) = 5 or N(adj)/N(act) = 5 & (16−R(adj))/(16−R(act)) = 7

From the definition, N must be the cumulative frequency, in decreasing order, at the particular last two throw totals required for Nia to reach the target. Likewise (16 - R) must be the particular cumulative frequency, but in increasing order, at which Rhys' last throw total takes him to just below the target.

Table 1 (one throw)			Table 2 (two throws)					
sum	frequency	cum. frequency	sum	frequency	cum. frequency	sum	frequency	cum. frequency
2	1	1	4	1	256	11	40	106
3	2	3	5	4	255	12	31	66
4	3	6	6	10	251	13	20	35
5	4	10	7	20	241	14	10	15
6	3	13	8	31	221	15	4	5
7	2	15	9	40	190	16	1	1
8	1	16	10	44	150		256	
	16							

Only on table 2 can the cumulative frequencies change by a factor of 7, therefore N(adj)/N(act) = 7. If the total of the second two throws reduces from 15 to 13 then the cumulative chance for this total and above increases from 5 to 35 ie 7 times. For the last two throw totals to be able to reduce by 2 then the first throw total needs to increase by 2.

As this is double this throw then the first throw was 2, which when added to the two second throws minimum total of 15 gives a target of 17.

Likewise if Rhys needed to score below a 7 on his last throw instead of actually needing to score below a 3 then Nia's chance of winning would increase by a factor of 5. Rhys' first two throw totals were therefore 17 minus 3 minus 1 (minus 1 to be below target) = 13.

Note 1 If Nia's chance increases by 7/3 and Rhys' chances are reduced by a factor of 15 then Nia's overall 35 times improvement is achieved, and these are from the tables achievable changes. However Nia's first throw total needs to be increased by only 1 to have a 7/3 increase. This is not possible for a doubling of her first throw and so this solution can be eliminated.

Answer: 16

There are five department shapes that have at least one axis of symmetry. The diagrams show the ways that the two symmetrical departments can be placed (up to reflection and rotation) that allows the four departments to be completed with four different shapes. We seek a solution in which all four departments have prime county totals.

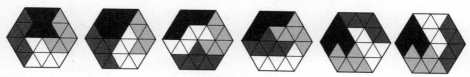

Diagram 1

We call the symmetrical shapes in Diagram 1 glider, pagoda, yacht, hexagon and dart. Opposite county numbers in any hexagon have equal total T, so the sum of the three opposite pairs is 3T and hence not prime.

Diagram 2   Diagram 3   Diagram 4

The pattern of odd (grey) and even (white) numbers (Diag 2) means that in the six positions a pagoda contains either two or four odd-numbered counties, so its county total is even and not prime.

A dart orientated North-South contains two or four odd-numbered counties and has an even total. (Diag 3).

Darts in other orientations can be seen as hexagons with one county translated (the circled counties in Diag 4). The difference between the original and translated county is always a multiple of 3, and the hexagon total is a multiple of 3 so the darts in these orientations have totals that are a multiple of 3 and are not prime.

So only the 3rd and 4th configurations in Diagram 1, involving a glider & yacht, might give four prime totals. The glider's total is a prime number in all six of its positions. In the 3rd configuration the boat has a prime total in four positions shown. In each case the glider has two positions and the completion is then forced; one of the departments has a composite total given. (Diagram 5)

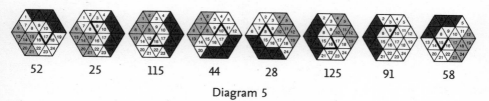

52   25   115   44   28   125   91   58

Diagram 5

In the 4th configuration the boat can be in two positions for each position of the glider and the forced completion is shown by the bold black border. A composite total is given for some department for except last case, where each total is prime.

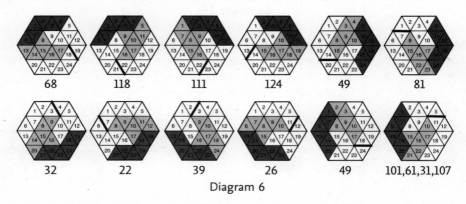

| 68 | 118 | 111 | 124 | 49 | 81 |

| 32 | 22 | 39 | 26 | 49 | 101,61,31,107 |

Diagram 6

The president's residence is built somewhere on the 6-13 border and the prime minister's mansion is in the middle of 16.

Answer: 15

Number of sides from 11, 23, 29, 41, 43, 47, 61, 67, 83, 89

Regular convex polygon only has whole-number internal angle, in degrees, for number of sides, N, being a factor of 360 [Total internal angle/N=180(N-2)/N=180-360/N], which is odd for N= 360, 120, **72**, **40**, 24, 8

Mean of four different primes above >=[11+23+29+41]/4=26 and <=[61+67+83+89]/4=75 So mean is from **40 or 72**. So sum of four **=160 or 288**

Eliminate 288: [89+83+67]=239 – so need +49 to reach 288 – invalid

So **mean=40, total=160** – last digit sequence from:-

_1+_1+29+89 or _3+_3+47+67 or _1+_3+_7+_9 or _1+23+43+83 or 11+41+61+_7

a1+b1+29+89=118+[11+41, 11+61 or 41+61]=170, 190 or 220 – all invalid

a3+b3+47+67=114+[23+43, 23+83 or 43+83]=180, 220 or 240 – all invalid

a1+23+43+83=149+[11, 41 or 61]=**160**, 190 or 210
but for 11+23+43+83=160, mean=40, $<x^2>-<x>^2$=747 – not square – invalid

11+41+61+d7=113+[47 or 67]=**160** or 180
but for 11+41+61+47=160, mean=40, $<x^2>-<x>^2$=333 – not square – invalid

a1+b3+c7+d9=160, so 11 invalid; 1+3+7+9=20; so a+b+c+d=14

Tabulate for a+b+c+d

a+c+d= b=	4+4+2	4+4+8	4+6+2	4+6+8	6+4+2	6+4+8	6+6+2	6+6+8
2	12	18	**14**	20	**14**	20	16	22
4	**14**	20	16	22	16	22	18	24
8	18	24	20	26	20	26	22	28

Primes	41,23,67,29				61,23,47,29			41,43,47,29
$<x^2>-<x>^2$	285				**225**			45
St. dev.	√285 invalid				**√225=15**			√45 invalid

Answer: 41

If there are n minor units in a major unit, and the side of the square is x major units and y minor units, then the puzzle requires a solution to $(xn+y)^2 = zn^2+15$ in non-negative integers, where $x<y<n \leq 10$.

Clearly $n^2>15$, or 15 couldn't be the remainder when the square major units are taken out. So n is between 4 and 10 inclusive.

If n is a multiple of 3, say $3r$, then the area is $(3rx+y)^2 = 9r^2x^2 + 6rxy + y^2 = 9r^2z + 15$. So 3 must also divide y, say $y=3s$. Then $9r^2(x^2-z) + 18rsx + 9s^2 = 15$. But the left hand side of this is divisible by 9 while the right hand side isn't, so this is impossible.

A similar argument shows that n can't be a multiple of 5.

If n is a multiple of 2, say $n=2r$, then $4r^2x^2 + 4rxy + y^2 = 4r^2z + 15$. Since 4 divides three of the terms, it must also divide $15-y^2$. So y must certainly be odd, say $y=2s+1$. Then 4 has to divide $15-4s^2-4s-1 = 14-4s^2-4s$, which is impossible because 4 does not divide 14.

So the only possibility for n is 7.

x cannot be zero; if it were, then the length of the side would just be y and the area would be y^2, which is certainly less than n^2, so we would need $y^2=15$, which is impossible.

y must be bigger than x and less than n, so it is between 2 & 6 inclusive. Look at all possibilities for $(xn+y)^2-15$ and see which are exactly divisible by n^2 (which should result in the value for z).

If $y=2$, $(7x+2)^2-15 = 49x^2 + 28x - 11$, which can't be divisible by 7, let alone 7^2.

If $y=3$, $(7x+3)^2-15 = 49x^2 + 42x - 6$, which can't be divisible by 7, let alone 7^2.

If $y=4$, $(7x+4)^2-15 = 49x^2 + 56x + 1$, which can't be divisible by 7, let alone 7^2.

If $y=5$, $(7x+5)^2-15 = 49x^2 + 70x + 10$, which can't be divisible by 7, let alone 7^2.

If $y=6$, $(7x+6)^2-15 = 49x^2 + 84x + 21$, which is divisible by 7 to give $7x^2 + 12x + 3$, which for x less than y can be divisible by another 7 only if $x = 5$, which gives a valid solution.

So the only values that work are $n=7$, $x=5$, $y=6$. Therefore the quantity asked for, the length of a side expressed just in minor units, is $5\times7+6 = 41$.

As a check, the area is $41^2 = 1681 = 34\times7^2+15$.

[Note that there is a more complex algorithm that finds all solutions for a general question of this type, but for the particular values of this Teaser there is the much simpler approach shown above.]

51 MAIN LINE

Answer: 65km

Consider the 'return route' from Boris's house (slow train to Anton's house and passenger train back to Boris's house).
The slow train takes 3/4 of the total time, the passenger train 1/4

Consider the 'return route' from Anton's house (goods train to Boris's house and passenger train back to Anton's house).
The goods train takes 3/5 of the total time, the passenger train 2/5

As the times for the two return routes are equal it follows that the ratio of the speeds of the two passenger trains is 2/5 : 1/4, i.e. 8 : 5

The passenger trains take an equal time from the passing point to the houses.
It follows that 25km represents 5/13 of the total distance between the houses.

The houses are **65km** apart.

52   BIRTHDAY MONEY

Answer: £6,660

If you ignore the missing year this is an Arithmetic Progression (AP).
The sum of an AP is the number of terms multiplied by the average of the first and last terms.
If 'n' is the age of my elder daughter then she will have received, ignoring the missing year, $n(5+5n)/2 = 5n(n+1)/2$ pounds in total.
Let x be her age when she received nothing, so her average birthday present was -
$\{5n(n+1)/2 - 5x\}/n = 5\{n(n+1)/2 - x\}/n$ (1)

The average birthday present for my younger daughter is therefore
$\{5(n-7)(n-7+1)/2 - 5(x-7)\}/(n-7) = 5\{(n-7)(n-7+1)/2 - (x-7)\}/(n-7)$ (2)
So, $5\{n(n+1)/2 - x\}/n = 1.21 * 5\{(n-7)(n-7+1)/2 - (x-7)\}/(n-7)$

This reduces to $x = n(21n^2 - 973n + 7476)/[14(3n+100)]$ (3)

But $1 \le x \le n$ which gives the following two inequalities –

i) $n(21n^2 - 973n + 7476)/[14(3n+100)] \le n$

As 'n' must be positive it can be divided into both sides without affecting the direction of the inequality, doing this and also deducting 1 from both sides gives –
$(21n^2 - 973n + 7476)/[14(3n+100)] - 1 \le 0$ or $(21n^2 - 1015n + 6076)/[14(3n+100)] \le 0$
Using the Quadratic Formula, $\{-b \pm \sqrt{(b^2-4ac)}\}/2a$, gives the following approx. factorisation
$21(n - 41.3)(n - 7)/[14(3n+100)] \le 0$, so $7 \le n \le 41$ (4)

ii) $n(21n^2 - 973n + 7476)/[14(3n+100)] \ge 1$ again this factorises to approx.

$21n(n - 9.7)(n - 36.6)/[14(3n+100)] \ge 1$, so $n \le 9$ or $n \ge 37$ (5)

As 'n' must be greater than 7 combining (4) & (5) gives $37 \le n \le 41$ or $8 \le n \le 9$

Using these values for 'n' in equation (3) only gives an integer value of 'x' of 28 when 'n' is equal to 40. (We are told that the daughters are grown-up so 'n' of 8 and 9 can be eliminated, these values also don't produce integer values of 'x'.)

Substituting n = 40 and x = 28 in equation (1), and without dividing by 'n', gives a total amount given to my elder daughter of £3,960. Doing the same in equation (2), and not dividing by (n - 7), gives a total amount given to my younger daughter of £2,700.

The total amount given to both daughters is therefore £6,660.

53 A CHRISTMAS CAROL

Answer: 29 February 1796

Marley always works on Christmas Day, so the date of the first announcement was 25/12/1835, which was Marley's last Christmas Day, since he died on 24/12/1836 and Christmas Carol was published almost seven years later on 19/12/1843.

Marley never works on his birthday, so in 1835 Marley worked between 1 and 364 days inclusive.

The announcement on 25/12/1835 relates to the Leap Year 1836.

If they had to work {1,2,3} dates then after levelling up to the average 2, they have to work {2,2,3} dates and no further fraction-less levelling is possible. {1,2,3} has the smallest maximum number that allows one round of fraction-less levelling up.

Let {x, 2,3} allow two rounds. After one round we have {1,2,3}. So {x+2+3}/3 = 1 and x=-2.

{-2,2,3} is not a permitted triple, but we can add 3 to get {1,5,6} (the minimum has to be 1 because they all work Christmas Day) as the triple with lowest maximum value that allows two rounds of fraction-less levelling up.

Repeating this process gives {1,2,3}, {1,5,6}, {1,14,15}, {1,41,42}, {1,122,123}, {1,365,366} as the allowable triples with lowest maximum, that allow 1, 2, 3, 4, 5, 6 rounds of levelling up.

Since 6 rounds of levelling up were made (on 26-31 December), the only possible allocation on 25/12/1835 was that in the following year the three men would work on {1,365,366} days.

Marley had worked between 1 and 364 days in 1835, so he must have been the character with only 1 work day in 1836 (Christmas Day) when Scrooge made his first announcement on Christmas Day.

After the first announcement Bob Cratchit was therefore down for 365 dates in 1836, the same dates he had worked in 1835, but had his birthday off in 1836, so his birthday must be February 29. So on Christmas Day the allocations are {Marley 1, Bob 365, Scrooge 366}.

1800 was not a leap year, but Cratchit is "in his 40s" on 19 December 1843. So he must have been born on 29 February 1796 making him 47 on publication day. Had he been born on 29 February 1804 his 40s would have begun on 29 February 1844. Had he been born on 29 February 1792 he would have been 51 on 19 December 1843.

54 SHOUT SNAP!

Answer: EIGHT SNAP THREE KING ACE SHOUT

Eight of the cards include a letter E and so those eight (shown on the left below) must take the

1st, 3rd, 5th, … and 15th positions. So the only adjacencies allowed are:

A; 2, 4, 6, K, SHOUT	**2**; A, 5, 7, 9, Q
3; 6, J, K, SNAP	**4**; A, 7, 8, 9, 10
5; 2, J, SHOUT, SNAP	**6**; A, 3, 10, Q
7; 2, 4, J	**J**; 3, 5, 7, 8, 9, 10, Q
8; 4, J, SNAP	**K**; A, 3
9; 2, 4, J, SHOUT	**SHOUT**; A, 5, 9
10; 4, 6, J	**SNAP**; 3, 5, 8
Q; 2, 6, J	

The K lies between the A and 3. To maintain the odd numbers in order, the 7 must come somewhere between the 5 and 9 and so SHOUT cannot lie between 5 and 9; i.e. it lies between the A and either 5 or 9. In order for the 5 to follow the 3 (with the 9 later) we must therefore have consecutive cards in order:
3 K A SHOUT 5

Now SNAP must lie between 8 and 3 or 5:

8 SNAP 3 K A SHOUT 5   or   3 K A SHOUT 5 SNAP 8

The 7 and 9 must follow this block and so the block of seven must be one of

The number of cards between the Q and J or K is even, and the number between the J and K is odd. Therefore the common number between 1st & 2nd and 2nd & 3rd picture-cards must be even, and it follows that the Q is the middle on the three picture-cards. The only fit is:

8	SN	3	K	A	SH	5		Q					J	

Only the 2 fits between the 5 and Q, so the 7 must then be between the 4 and J:

8	SN	3	K	A	SH	5	2	Q			4	7	J	9

That leads uniquely to:

8	SN	3	K	A	SH	5	2	Q	6	10	4	7	J	9

55   THERE'S ALWAYS NEXT YEAR

Answer: 500 and 165

The entry can be anything from 475 to 525 inclusive and the first point to realise is that, as the human being is indivisible, if the above percentages were to be precise, the number of entrants would have to be a multiple of 5 to have any chance at all and therefore only the following are possible:

Number of entrants	Percentage pass rate	Number of successes
475	32	152
	36	171
480	30	144
	35	168
490	30	147
500	30	150
	31	155
	32	160
	33	165
	34	170
	35	175
	36	180
510	30	153
520	30	156
	35	182
525	32	168
	36	189

Listing the percentage pass rates with the appropriate number of successful candidates:
30: 144 147 150 153 156
32: 152 160 168
35: 168 175 182
36: 171 180 189

Bearing in mind that the new pass percentage will have to be different, i.e. 31, 33 or 34, the only possibility is for the new entry list to number exactly 500.
Thus, with the four percentage pass rates already shown, 500 is ruled out and we have the following possibilities over the first four years:

Percentage pass rate	Number of entrants				Number of successes			
30	480	490	510	520	144	147	153	156
32	475	525			152	168		
35	480	520			168	182		
36	475	525			171	189		

Bearing in mind that the number of entrants and successes each year has to be different, the following are possible:

144 + 152 + 182 + 189 = 667
144 + 168 + 182 + 171 = 665
147 + 152 + 168 + 189 = 656
147 + 152 + 182 + 189 = 670
147 + 168 + 182 + 171 = 668
153 + 152 + 168 + 189 = 662
153 + 152 + 182 + 189 = 676
153 + 168 + 182 + 171 = 674
156 + 152 + 168 + 189 = 665

We can see by inspection that the number of successes lies between 144 and 189 and thus the only perfect square we can hope to hit by adding from the mid six-hundreds is 841. In order to get a whole number percentage pass rate, we need our new number to be exactly divisible by 5 and thus we must be adding a number ending in 1 (0 + 1 = 1) or 6 (5 + 6 = 11). We have two possibilities.

841 - 656 = 185 for a pass rate of 37%, which is outside the range.

841 - 676 = 165 for a pass rate of 33%, which is within the range.

56 DIDDUMS!

Answer: 8

If the prime factorization of $N=(p^a)(q^b)(r^c)...$ then N has $D=(a+1)(b+1)(c+1)...$ divisors, including 1 and N.

For D=number of digits of N=digit sum of N the following forms for N apply

D=1x	1	2		3	4	5	6	7	8	9
or					$2x2$		$2x3$		$2x4$	$3x3$
or									$2x2x2$	
N=	$p^0=1$	$p^1=11$		p^2	p^3	p^4	p^5	p^6	p^7	p^8
or					pq		pq^2		pq^3	$(p^2)(q^2)$
or									pqr	

D	N
3	digit sum=3, so N is div. by 3, so $p^2=3^2=9$ – invalid
4	possible numbers: 1003, 1012, 1021, 1030, 1102, 1111, 1120, 1201, 1210, 1300, 2002, 2011, 2020, 2101, 2110, 2200, 3001, 3010, 3100, 4000 None=p^3 ($11^3=1331$, $13^3=2197$). All ending '0'=2x5xM (not=pq) **1003=pq=17.59**, 1012=4M=(2^2)M, 1021=p, 1102=pqr=2.19.29, **1111=pq=11.101**, 1201=p, 2002=pqrs=2.11.7.13, 2011=p, **2101=pq=11.191**, 3001=p

So first five diddums are 1, 11, 1003, 1111, 2101

We are told that the sixth diddum is even

5	p^4 is only even for $2^4=16$ - invalid
6	digit sum=6, N div by 3, but not 9, so only even for $pq^2=3x2^2=12$ - invalid
7	p^6 is only even for $2^6=64$ - invalid
9	digit sum=9, N div by 9, so only even for $(p^2)(q^2)=3^2x2^2=36$ – invalid

For sixth diddum to be listed and even, the only remaining options have digit sum=8

8	possible numbers with digit-sum=8:

10000007=odd
10000016=16x625001=(p^4)M
10000025=odd - all invalid
10000034=pqr=2x11x454547 (after lengthy calculations on 454547) - OK

57 THAT STRIKES A CHORD

Answer: KNO

The allowable chord changes and the relevant chords are as shown in this table; the first column has notes in alphabetical order, while the other columns have the notes after their shifts, and these notes might not be in order. The strikethroughs indicate changes prohibited by the rules (one because it gives an unchanged chord, and the rest because they give repeated notes). The allowable chord changes are also shown in the diagram.

chord notes in	go to these after these shifts		
alphabetical order	+1, +1, -2	+1, -2, +1	-2, +1, +1
JNP	KON	KLJ	~~OOJ~~
KNO	LOM	~~LLP~~	~~POP~~
JKL	~~KLJ~~	KPM	OLM
KMP	~~LNN~~	LKJ	PNJ
LMO	~~MNM~~	MKP	JNP

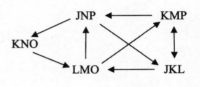

This shows that there are 5 different chords in the sequence, and 9 allowable transitions between chords. One chord (KNO) can only be followed by one other, but the remaining four of the chords can each be followed by two others. For these, each must follow at some stage of the sequence, and either of them could happen first, and therefore there can be at most $2^4 = 16$ sequences. Half of these will have the follower of the initial JNP being JKL; we can look at these first, and then find the other half by 'rotating' the sequence to bring the second JNP to the start.

This gives the following possibilities. In three cases JNP occurs for a third time before the end of the sequence, and then there is no allowable chord left to follow it. The table also shows the number of chord changes between successive occurrences of JNP.

sequence of chords										changes
1	2	3	4	5	6	7	8	9	10	between JNPs
JNP	JKL	KMP	JKL	LMO	JNP	KNO	LMO	KMP	JNP	5 & 4
JNP	JKL	KMP	JKL	LMO	KMP	JNP	KNO	LMO	JNP	6 & 3
JNP	JKL	KMP	JNP	KNO	LMO	JNP	none left			n/a
JNP	JKL	KMP	JNP	KNO	LMO	KMP	JKL	LMO	JNP	3 & 6
JNP	JKL	LMO	JNP	KNO	LMO	KMP	JKL	KMP	JNP	3 & 6
JNP	JKL	LMO	JNP	KNO	LMO	KMP	JNP	none left		n/a
JNP	JKL	LMO	KMP	JKL	KMP	JNP	KNO	LMO	JNP	6 & 3
JNP	JKL	LMO	KMP	JNP	KNO	LMO	JNP	none left		n/a

The candidate solutions that have more changes between the first pair of JNPs than between the second pair of JNPs would be: that in row 1 as it stands, with 5 as opposed to 4; those in rows 2 & 7 as they stand, with 6 as opposed to 3; and those in rows 4 & 5, but starting at column 4 and winding round, also with 6 as opposed to 3. The unique candidate solution that gives the first gap as small as it can be is the one in row 1. The seventh chord of this is KNO.

58 CRAZY GOLF

Answer: £32

Ian won hole 1 against John and hole 9 against Ken and only 3 other holes, so he won 3 holes against one of them, and 2 holes against the other in total.

Winning three holes out of 9, losing sequences are:

Sequence	Cost	Total
6	£21	£21
5 + 1	£15 + £1	£16
4 + 2	£10 + £3	£13
3 + 3	£6 + £6	£12
4 + 1 + 1	£10 + £1 + £1	£12
3 + 2 + 1	£6 + £3 +£1	£10
2 + 2 +2	3 x £3	£9

Winning two holes out of nine:

Sequence	Cost	Total
7	£28	£28
6 + 1	£21 + £1	£22
5 + 2	£15 + £3	£18
4 + 3	£10 + £6	£16

The only way in which both boys received equal money was £16 + £16

So, Ian paid out £32

59 WONKY DICE

Answer: 1 2 2 3 3 4 4 5

The probability of a particular total is the number of ways of obtaining that total divided by 64. For a standard pair of octahedral dice, the scores and ways are:

Score	2	3	4	5	6	7	8	9	10	11	12	13	14	15	16
Ways	1	2	3	4	5	6	7	8	7	6	5	4	3	2	1

For the wonky dice, there must be a single 1 on each die to give a 1/64 probability of scoring 2.

If 11 is on the first die then logically the highest number on the second die is 5, and there can only be one 11 on the first die, and only one 5 on the second die. This gives a 1/64 probability of scoring 16.

Considering the second die, there must be no more than two 2s (otherwise the probability of a total of 3 is more than 2/64), no more than three 3s (otherwise the probability of a total of 4 is more than 3/64), and no more than two 4s (otherwise the probability of 15 is more than 2/64). Hence the numbers on the second die are:

(a) 1 2 3 3 3 4 4 5
(b) 1 2 2 3 3 4 4 5 or
(c) 1 2 2 3 3 3 4 5

Now consider the second-highest number on the first die.

If it is 10, then (a) and (b) are ruled out because there are more than three ways of obtaining a total of 14 and (c) is ruled out because there are more than 4 ways of obtaining a total of 13.

If it is 8 or less, then only (a) is possible (to get three ways of obtaining 14 and two ways of obtaining 15). Only 8 will allow four ways of obtaining 13, and only if there are three 8s, but this gives too many ways of obtaining 12.

Therefore the second-highest number on the first die is 9, and (b) is the only possibility (to give three ways of obtaining 14 and two ways of obtaining 15).

For completeness, the numbers on the first die are: 1 3 5 5 7 7 9 11

60 RECURRING THEME

Answer: 4, 30, 40 and 74

If the reciprocal of a number does not give rise to a recurring decimal then its multiples cannot do so. The first step is therefore to take the reciprocals of numbers 7, 17.......97 and 70, 71......79. This shows that the only possible values for the largest number are 27, 37 and 74 (see Appendix 1).

It is sensible to first check 10 and 20 as numerators with 27 as the denominator, 10, 20 and 30 with 37, and also 10 to 70 with 74. There will be no unique cases involving no or one number divisible by ten (see Appendix 2).

	37			74			27
10	0.270...		10	0.135...		10	0.370...
20	0.540...		20	0.270...		20	0.740...
30	0.810..		30*	0.405...			
			40*	0.540...			
			50	0.675...			
			60	0.810...			
			70	0.945...			

The only unique case arises if 74 is the denominator and two of the numerators are 30 and 40. The other numerator is found by using the digits 0, 4 and 5 to generate other recurring decimals and multiplying them by 74. This leads to .054054... x 74 = 4.

The four numbers are 4, 30, 40 and 74.

Appendix 1

7	17	27	37	47
0.142857	0.058823	0.037037	0.027027	0.021276
57	67	77	87	97
0.017543	0.014925	0.012987	0.011494	0.010309
70	71	72	73	74
0.014285	0.014084	0.013888	0.013698	0.0135135
75	76	77	78	79
0.013333	0.013157	0.012987	0.012820	0.012658

Appendix 2 It is only necessary to consider enough examples to show that there is more than one case with no or one numbers divisible by ten.

37	10			1	26
	0.207...	0.027...	0.702...		
37	20			2	15
	0.540...	0.054...	0.405...		
37	5			13	19
	0.135...	0.351...	0.513...		
37	6			8	23
	0.162...	0.216...	0.621...		

Answer: 1815

Let the dimensions of the ingot be positive integers x,y,z with 1< x < y <z. Then V = xyz.
V + S + E = xyz + (2xy + 2yz +2zx) + (4x + 4y + 4z) = (x + 2)(y + 2)(z + 2) − 8
Estella's value V+S+E is 3177 so 3177 = (x + 2)(y + 2)(z + 2) − 8 and (x + 2)(y + 2)(z + 2)
= 3185 = $5 \times 7^2 \times 13$ which does not uniquely factorise into the product of three different
integers, so Estella does not know (x+2,y+2,z+2) and hence Pip's value. Since (x+2) can't
be 1, the possibilities from Estella's point of view are:

Estella: (x+2, y+2, z+2, V+S+E+8) a] (5, 13, 49, 3185)   b] (7, 13, 35, 3185)   c] (5, 7, 91, 3185)

Pip: (x, y, z, V)   a] (3, 11, 47, 3×11×47) b] (5, 11, 33, 5×11×33) c] (3, 5, 89, 3×5×89)

If case a] or c] pertained then Pip would claim at the first bell because his value
factorises uniquely into the product of three primes and because he is clever would have
no difficulty in discovering this factorisation and Estella's value. If case b] pertains Pip
does not know which factorisation of V is correct (it could be 5×11×33, 3×5×121 or
3×11×55) and hence cannot work out Estella's value and claim at the bell.

So when Pip doesn't claim, Estella knows that case b] pertains. All she has to do in the
next fifteen minutes is calculate Pip's value 5×11×33 = 1815 and she can write 3177,
1815 on the crate when the bell rings.

Now we have to check (given Estella's claim) that Pip cannot use similar logic.
Given Estella's claim, Pip has value 1815 and the possibilities from Pip's point of view
(recall x > 1) are

Estella: (x+2, y+2, z+2, V+S+E+8) a] (5, 7, 123, 5×7×123) b] (7, 13, 35, 7×13×35) c] (5, 13, 57, 5×13×57)

Pip: (x, y, z, V)   a] (3, 5, 121, 1815)   b] (5, 11, 33, 1815)   c] (3, 11, 55, 1815)

As Pip sees it, Estella cannot make a claim at first bell if a], b] or c] pertain because in
each case her V+S+E + 8 does not factorise uniquely as a product of three integers (123,
35, and 57 are not prime). Pip cannot rule out any of his cases just because Estella did
not claim at first bell.

[In *Great Expectations* by Charles Dickens, Estella does indeed get the bulk of Miss
Havisham's will and Pip nothing. In the problem Miss Havisham fixes this by making
sure Estella is given the V+S+E envelope, typical of her mind games.]

Answer: 7

Let 'n' be the number of laps cycled by Alan. Let 'k' be the number of laps walked by Bob. Let 'R' be the sum raised by each.

Therefore:

For Alan, $R = 1 + 2 + 3 + \ldots + n = n(n+1)/2$

For Bob, $R = 1 + 2 + 4 + \ldots + 2^{k-1} = 2^k - 1$

Given R is a four figure sum we know: $1,000 <= 2^k - 1 <= 9,999$. Therefore, $10 <= k <= 13$.

Equivalence: $n(n+1)/2 = 2^k - 1$. Therefore, $n = [(2^{k+3} - 7)^{0.5} - 1]/2$. So we require solutions of form $2^{k+3} - 7 = x^2$ for an integer x.

Given constraints on k:

$k = 10$: $2^{10+3} - 7 = 8,185$.	$90^2 < 8,185 < 91^2$.
$k = 11$: $2^{11+3} - 7 = 16,377$.	$127^2 < 16,377 < 128^2$.
$k = 12$: $2^{12+3} - 7 = 32,761$.	$32761 = 181^2$.
$k = 13$: $2^{13+3} - 7 = 65,529$.	$255^2 < 65,529 < 256^2$.

Therefore, $k = 12$ is the only candidate, and $n = [181 - 1]/2 = 90$.

Therefore Alan completes $n = 90$ laps, and Bob completes $k = 12$ laps, and each raise $R = £4,095$.

The greatest common divisor of 90 and 12 is 6. Therefore, given constant speeds, they cross the start-finish line together $1 + 6 = 7$ times (adding one for the start). This can be visualised as follows:

Laps Completed

Alan	0	15	30	45	60	75	90
Bob	0	2	4	6	8	10	12
intersection	1	2	3	4	5	6	7

63 JOKERS

Answer: 15

Regarding the cards as 52 of one type and J of another type there are $(52 + J)!/[52!J!]$ ways of ordering the cards. Avoiding two jokers together is like placing the J jokers in the 53 'gaps' (including the ends) between the 52 cards and this can happen in $53!/[(53 - J)!J!]$ ways. So the probability of <u>avoiding</u> two jokers together is the second expression divided by the first which simplifies to
$$53/(52 + J) \times 52/(51 + J) \times 51/(50 + J) \times \ldots \text{(J terms)}$$

For J=6 and 7 this gives
J=6:   53.52.51.50.49.48/58.57.56.55.54.53 ≈ 0.57
J=7: 0.57 × 47/59 ≈ 0.45

So for there to be a more than 50:50 chance of two jokers being together we must have J≥7.

Now $52 + J$ is the smallest number of 52 or more that is divisible by the number of players, P. So P is not a factor of 52, 53, 54, 55, 56, 57 or 58. Also P is at most 17 because otherwise 52/P is less than 3 and once the J cards are added each player will get at most 3 cards. That just leaves the following possibilities for P, J and the number of cards each, C:

P	10	12	**15**	16	17
J	8	8	**8**	12	16
C	6	5	**4**	4	4

The probability of there being no joker in my C cards is
$$52/(52 + J) \times 51/(51 + J) \times 50/(50 + J) \times \ldots \text{(C terms)}$$

The obvious case that maximises this expression is P=15, J=8, C=4 which gives a probability of
52.51.50.49/60.59.58.57 ≈ 0.56

The cases to the left and right of that both give answers of less than 0.5, so **P=15** is the only case where there is more than a 50:50 chance of me having no jokers.

Answer: 127 feet

Let A be the start of the cable and D be its end, with BC being the ducted length over the exposed area KLNM. x and y are the sides of the garden, with x < y.

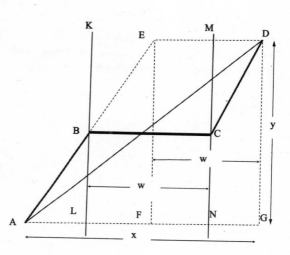

To minimize the length of ducting, angles LBC and NCB need to be 90°.

Let the width of the exposed area be w and let E be a point at right angles to DG and at the same width w from D.

The total cable length is therefore AB plus CD plus BC. As BCDE is a parallelogram, CD = BE. So the total cable length is AB plus BE plus ED.

This distance is minimized when AE is a straight line. From the diagram $AD^2 = x^2 + y^2 = 123^2 = (3*41)^2$

This is a Pythagoras Triple $(3(a, b, c))$, and from the box below $= 3(m^2 - n^2, 2mn, m^2 + n^2)$. Note it has to be a multiple of 3 as $m^2 + n^2 \neq 123$. So $m^2 + n^2 = 41$ and from observation $m = 5$ and $n = 4$ so that the Triple is $(9, 40, 41)$ and multiplied by 3 is $(27, 120, 123)$. So $x = 27$ and $y = 120$.

The cable length is AE + w, as AE must be an integer then $\triangle$ AEF produces another Pythagoras Triple of size $(27 - w, y, AE)$. From the box $y = k*2mn = 120$. $kmn = 60$. Also $(27 - w) = k(m^2 - n^2)$, this must be less than 27 which is only possible when k and the difference between m and n are both small, with the smallest combinations, for low values of k being –

k	m	n	$k(m^2 - n^2)$		
1	10	6	64		
				Only valid solution.	
2	6	5	22	Check k*2mn = 120	No other combinations of k, m and n give valid solutions
3	5	4	27		
4	5	3	64		
5	4	3	35		

The Triple is therefore 2 (11, 60, 61) = (22, 120, 122). Therefore AE =122 and w = 5.

The minimum length of cable is therefore AE + w = 122 + 5 = 127 feet

Formula for finding Pythagoras Triples

a, b and c in the right-angled triangle are integers.

Let $m^2 - n^2 = a$ and $2mn = b$ where m & n are integers and $m > n > 0$

Then $(m^2 - n^2)^2 + (2mn)^2 = (m^2 + n^2)^2$ and as $a^2 + b^2 = c^2$,

$c = m^2 + n^2$

The Triple is then $(m^2 - n^2, 2mn, m^2 + n^2)$

And with all Triples, including non-primitive, being $k*(m^2 - n^2, 2mn, m^2 + n^2)$ where k is an integer > 0

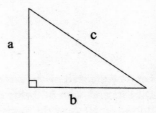

Answer: 18, 59

Let there be N clubs in the league, so N>13. In a season each club plays 2*(N-1) matches, and the total number of all matches in the league is N*(N-1). There are 21 double-digit prime numbers, namely 11, 13, 17, 19, 23, 29, 31, 37, 41, 43, 47, 53, 59, 61, 67, 71, 73, 79, 83, 89 and 97, so N<22. We now tabulate the possible results for Wessex Wanderers (WW) as follows:

Losses	1	2	3	4	5	6	7	8	9	10
Wins	3	6	9	12	15	18	21	24	27	30
Draws	2N-6	2N-10	2N-14	2N-18	2N-22	2N-26	2N-30	2N-34	2N-38	2N-42
Points	2N+3	2N+8	2N+13	2N+18	2N+23	2N+28	2N+33	2N+38	2N+43	2N+48

Since N<22, larger numbers of losses would result in negative numbers of draws. We note that, for any value of N, the points totals 2N+8, 2N+18, 2N+28, 2N+38 and 2N+48 are always even and cannot be prime numbers. If N is 16 or 21, the corresponding points totals are all multiples of 5, and again cannot be prime numbers. So we are left to consider the values of N equal to 14, 15, 17, 18, 19, and 20 for the following points totals:

N	2N+3	2N+13	2N+23	2N+33	2N+43
14	<u>31</u>	<u>41</u>	51	<u>61</u>*	<u>71</u>*
15	33	<u>43</u>	<u>53</u>	63	<u>73</u>*
17	<u>37</u>	<u>47</u>	57	<u>67</u>	77*
18	39	49	<u>59</u>	69	<u>79</u>*
19	<u>41</u>	51	<u>61</u>	<u>71</u>	81*
20	43	53	63	<u>73</u>	<u>83</u>

The entries in this table marked by asterisks are invalid because they all arise from negative numbers of drawn matches. The entries that are underlined are all double-digit prime numbers. The only row in the table with only one unstarred double-digit prime number is for N=18, with the double-digit prime number 59. So there are 18 clubs in the league, and the final points total for WW is 59.

For completeness, we need to show that a final league table including the above outcome for Wessex Wanderers is actually possible, and the following table shows that it is:

Team	Wins	Draws	Losses	Points
1	23	10	1	79
2	20	13	1	73
3	19	14	1	71
4	17	16	1	67
5	15	16	3	61
6 (WW)	15	14	5	59
7	15	8	11	53
8	14	5	15	47
9	5	28	1	43
10	4	29	1	41
11	3	28	3	37
12	4	19	11	31
13	0	29	5	29
14	2	17	15	23
15	1	16	17	19
16	2	11	21	17
17	2	7	25	13
18	1	8	25	11
Totals	162	288	162	774

66 ROMAN PRIMES

Answer: 5, 37 and 97

There are seven possible sets of characters on the die faces:-

Case 1	Case 2	Case 3	Case 4	Case 5	Case 6	Case 7
IVXLCD	IVXLCM	IVXLDM	IVXCDM	IVLCDM	IXLCDM	VXLCDM

In cases 1 and 2 the possible prime sequences are as follows:-

II (2) III (3) XIII (13) XXIII (23) LXXIII (73) LXXXIII (83)
II (2) III (3) XIII (13) XLIII (43) LXXIII (73) LXXXIII (83)
II (2) III (3) LIII (53) XLIII (43) LXXIII (73) LXXXIII (83)
II (2) VII (7) XVII (17) XLVII (47)
II (2) VII (7) XVII (17) LXVII (67)
II (2) VII (7) XVII (17) XCVII (97)
XI (11) XIX (19) XXIX (29) LXXIX (79) LXXXIX (89)
XI (11) XIX (19) XXXI (31) LXXIX (79) LXXXIX (89)
XI (11) XIX (19) LXXI (71) LXXIX (79) LXXXIX (89)
XI (11) XLI (41) LXXI (71) LXXIX (79) LXXXIX (89)
XI (11) LIX (59) LXXI (71) LXXIX (79) LXXXIX (89)
XI (11) LXI (61) LXXI (71) LXXIX (79) LXXXIX (89)

Every prime between 1 and 100 appears in at least one of these sequences, except for 5 and 37. It is not possible to obtain 5 = V because every sequence starts by using two characters to form the first prime. Even though 37 = XXXVII, which can be formed using the characters on the dice, it is not possible to obtain XXXVII by adding a character to one of the five-character primes and rearranging the six characters.

In case 3, the possible prime sequences are the same except for the sixth one. Character C is missing in this case so 97 cannot be formed and there are three primes (5, 37 and 97) which do not appear in any prime sequence.

In cases 4 to 7, characters L, X, V or I are missing so there are more than three primes less than 100 which cannot be formed and which do not appear in any prime sequence.

Case 3 is the only one in which exactly three primes are missing and these are 5, 37 and 97.

67 UNDERGROUND UMBRA CONUNDRUM

Answer: 8/15

For a linear shadow edge to only hit two intersections at four tile corner points with the symmetry described, the offset between the upper and lower 'levels' must be even >6 <=18. Additionally the 'edge slope'=vertical offset between points/horizontal offset between points must be a fraction in simplest form (and >20/15=4/3 else shadow crosses sides of the wall).

Shadow edges cross every colour – so horizontal offset>1

Vert. Offset V=	8	10	12	14	16	18
Hor. Offset H=	3; 5	3; 7	5; 7	3; 5; 9	3; 5; 7; 9; 11	5; 7; 11; 13

For V=8, 10, 12 and 18 – no permissible shadow results with a column completely in shadow.

For V=14: 14/9 can't fit with either 14/5 or 14/3 to give a full-shadow column. 14/5 and 14/3 can fit grid to match conditions as shown below in the left-hand diagram. Edge AC (slope 14/5) just cuts into 1st and 9th columns, since 14 up or down and 5 along = 3 up or down and 15/14 along. Area of shadow = 20x(AB+CD)/2 = 20x(7+9)/2 = 20x8. Wall area = 20x15, so area ratio is 8/15.

For V=16: 16/9 and 16/11 can't fit with any other to give full-shadow column. Only 16/7 and 16/3 can fit grid to match conditions as in the right-hand diagram. Edge AC (slope 16/7) lies across 1st and 9th columns only, since 16 up or down and 7 along = 2 up or down and 7/8 along. Area of shadow = 20x(AB+CD)/2 = 20x(6+10)/2 = 20x8. Wall area = 20x15, so area ratio is 8/15.

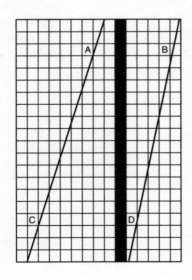 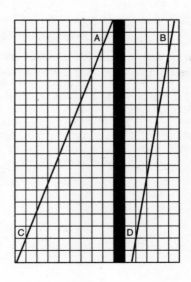

68 ROOM 101

Answer: 15625, 34969, 62001 and 91809

$1984 = 2^6.31$ so Winston's squares are odd and not divisible by 31. Let the 5-digit squares Winston chooses be $a_1^2, a_2^2, a_3^2, ..., a_n^2$, $n > 50$, $100 \le a_i \le 316$ (5-digits)

Let their remainders after division by 101, be $r_1, r_2, r_3, ..., r_n$, $0 \le r_i \le 100$

No two remainders can be the same because if $r_i = r_j$ then $a_i - a_j$ is divisible by 101 and $a_i^2 - a_j^2 = (a_i - a_j)(a_i + a_j)$ would be divisible by 101.

Similarly the sum of any two remainders $r_i + r_j$ cannot be divisible by 101 because then $a_i + a_j$ would be divisible by 101 and so would $a_i^2 - a_j^2 = (a_i - a_j)(a_i + a_j)$.

Because 101 is prime the only way for $a_i^2 - a_j^2 = (a_i - a_j)(a_i + a_j)$ to be divisible by 101 is for one or other factor to be divisible by 101, so provided Winston chooses squares for which $r_i \ne r_j$ and $r_i + r_j$ is not divisible by 101, and none of them are divisible by 2 or 31 he will meet O'Brien's requirements.

Winston must necessarily choose only one remainder from the fifty pairs (1, 100), (2, 99), (3, 98), ... (50, 51). He can also choose either 101^2 or 303^2 to get "more than fifty" but no others. Winston tells O'Brien he didn't choose 101^2 so O'Brien knows he chose $303^2 =$ **91809**.

In choosing from remainders (1, 100) the only non-even a_i are 202+1 = 203 or 101+100 = 201; Winston can choose either of those because neither is divisible by 31. O'Brien does not know which of those Winston chose.

In choosing from remainders (2, 99) the only non-even a_i are 101+2 = 103 or 303+2 = 305 or 99+202 = 301; Winston can choose any of these because none are divisible by 31. O'Brien does not know which of those Winston chose.

In general if the remainder pairs are (odd, even) then Winston can usually choose between two odd a_i, namely 202+ odd or 101 + even except when one of these is divisible by 31. If the remainder pairs are (even, odd) then Winston can usually choose between two odd a_i, namely   101+ even or 202 + odd (and 303 + even, until this exceeds 316) except when one of these is divisible by 31.

So for each pair $(r, 101–r)$ there is always a choice of at least two odd . The pinch points are when potential a_i are odd but divisible by 31. These multiples of 31 are 155, 217, 279. The relevant remainder pairs are (15, 86), (24, 77), (47, 54).

In the case of (15, 86) the odd a_i are 202 + 15 = 217 or 86 + 101 = 187. Now Winston has no choice because 217^2 has factor 31 shared with 1984. He must choose $187^2 =$ **34969**.

In the case of (24, 77) the odd a_i are 101 + 24 = 125 or 77 + 202 = 279. Now Winston has no choice because 279^2 has factor 31 shared with 1984. He must choose $125^2 =$ **15625**.

In the case of (47, 54) the odd a_i are 202 + 47 = 249 or 54 + 101 = 155. Now Winston has no choice because 155^2 has factor 31 shared with 1984. He must choose $249^2 =$ **62001**.

So O'Brien knows that among the quadrillion ways of choosing the 51 squares, Winston must choose 15625, 34969, 62001 and 91809.

We can justify Winston's "at least a quadrillion" by observing that for remainder pairs (2, 99), (4, 97), (6, 95), ... (12, 89) Winston has three choices he can make (finishing with 113 or 315 or 291). Otherwise, he has two choices, except in the three cases given where he has only one choice; and he must choose 91809. Winston can therefore choose his 51 squares in $3^6.2^{41}.13 .1^1 > 1.6 \times 10^{15}$ ways which exceeds a quadrillion.

69 MY GRANDFATHER'S COINS

Answer: 311

First, let us determine N, the number of children.

Because the eldest received 1/5 of the coins, and each of the remaining N − 1 at least 1/ 11 of the coins we have: $1/5+(N-1)/11<1$

Hence, since N is an integer, N<10.

On the other hand, because the eldest received 1/5 of the coins, the second at most 1/5, and each of the remaining N − 2 at most 1/10 of the coins we have:

$1/5 + 1/5 + (N-2)/10 > 1$

Hence N>8.

Thus, the number of children is 9.

Now, if X is the number of coins the grandfather left, then X must be a multiple of 110, that is, one of the 22 elements of the set {110, 220, 330, ..., 2420}. However, between X/11 and X/10 we must fit at least 5 primes, and this is only possible if X = 2420, there being 6 primes between 220 and 242, namely: 223, 227, 229, 233, 239, and 241. The only choice of five of these which allows for Isabel's number of coins to be a prime (and the total number of coins to be 2420) is all six but the prime 229. In such a case this means Isabel received 311 coins.

Answer: 3, 17 & 35 and 5, 7 & 9

Given any 'hole length' s and three 'clubs' p, q & r, we need to solve $ap + bq + cr = s$ in integers, and then find the smallest sum of absolute values of a, b and c. This could be approached by methodical trial and error to find a particular solution, and then adjusted to minimise the sum of absolute values. Adjustments can be done by adding kq to a while subtracting kp from b, for any integer k (positive, negative or zero); or, similarly, adding mr to b while subtracting mq from c; or adding np to c while subtracting nr from a; or any combination of these. All of these clearly leave the total s unchanged.

A more orderly approach can be used, as described below. The approach might sound elaborate at first, but is quite simple to apply, and it is a way of showing uniqueness of solution.

Note first that no two of the club values in the Teaser have a common factor; this makes the following description easier, because otherwise we would be worrying about highest common factors other than 1, and the chance of there being no possible solutions.

The solution of the equation in three variables is simplified if it is split into two equations in two variables each, as $ap + d{\times}1 = s$ and $bq + cr = d$. We can find solutions for the first of these, and then substitute the value(s) for d in the second one.

Now, note that if we take any two solutions to, say, the equation $xu + yv = w$ in two variables (x & y), we get something like $(x_1-x_2)u + (y_1-y_2)v = 0$, and the only solutions to this are $x_1-x_2 = kv$, $y_1-y_2 = -ku$, for any integer k (positive, negative or zero) if u & v have no common factors. So from any particular solution we can find all possible solutions.

One efficient way to find a particular solution to $xu + yv = w$ is to consider remainders on dividing by u. Obviously xu leaves no remainder, and one can step through (the remainders of) successive multiples of v until one of them matches that of w, and this will be the yth. Then x can be found by substituting y into the equation. Or one can use the same approach to solve $xu + yv = 1$, and then multiply x & y by w to solve $xu + yv = w$; this can be used even if w is being expressed in terms of a parameter k as above, for instance as {particular solution} + kx{increment}.

Combining all the above leads to a completely general solution in the form $a = a_0 - \alpha$, $b = b_0(d_0+\alpha p) - \beta r$, $c = c_0(d_0+\alpha p) + \beta q$, for parameters α & β. Varying the parameters produces a two-dimensional grid of solutions. Linear combinations using steps of α & β can be formed, to find small grid elements. Then one can check when a minimal solution has been found for a hole.

There are 20 ways of choosing 3 clubs from the 5 available (5 ways for the first club, times 4 for the second, times 3 for the third, divided by 3 and 2 because the order of choosing is irrelevant). Without using the largest club, 35, the total score would be at least the value found by dividing each of the three hole lengths by the second largest club (19), rounding up nonintegral results to the next integer, and summing them, to get 24 strokes, so to achieve fewer than 24 strokes, the 35 must be used.

The table below shows the minimal results for all club selections that use 35. The coefficients show, for instance in the second column of the penultimate row, that $-1{\times}3 + 2{\times}17 + 2{\times}35 = 101$. The 'small grid steps' often show up as differences in alternative possible sets of coefficients. It can be checked that the coefficients shown give minimal numbers of strokes, because adding various combinations of grid steps cannot give a lower number.

chosen clubs	coefficients to reach hole lengths			small grid steps		total score
	101	151	197	one way	other way	
17 19 35	2 -2 3 or 5 -1 1	-1 7 1	0 3 4	8 -9 1	3 1 -2	7 + 9 + 7 = 23
8 19 35	-4 7 0 or -2 8 -1 or 0 9 -2	-1 1 4 or 1 2 3	0 3 4	-11 12 -4	2 1 -1	11 + 6 + 7 = 24
3 19 35	4 1 2 or 3 3 1 or 2 5 0	9 1 3 or 8 3 2 or 7 5 1 or 6 7 0 or -8 0 5	1 1 5 or 0 3 4	12 11 -7	-1 2 -1	7 + 13 + 7 = 27
8 17 35	4 2 1 or 2 5 0	-3 0 5	5 1 4 or 3 4 3 or 1 7 2	7 7 -5	-2 3 1	7 + 8 + 10 = 25
3 17 35	-1 2 2	-2 1 4	2 3 4	6 1 -1	5 -5 2	5 + 7 + 9 = 21
3 8 35	-4 1 3	1 1 4	1 -2 6 or 2 2 5	7 -7 1	1 4 -1	8 + 6 + 9 = 23

The lowest possible total of 21 requires clubs 3, 17 & 35, with individual hole scores of 5, 7 & 9.

71   MANY A SLIP

Answer: 826, 574 and 536

(a)
If the shaded letters are correct then C>A>L and U≠N. So if you could find a solution
CU? + AN? + L?? it follows that CN? + AU? + L?? would be another solution. So in these
cases you could not work out my three numbers.

(b)
If the shaded letters here are correct then we would require wxP + yzD to be 1000. If we
found such a solution we'd have x+z=9 and so x≠z and (as before) wzP + yxD would be a
different solution. Hence, once again, in these cases you could not work out my numbers.

(c) It follows from (a) and (b) that the incorrect letter is in both the shaded areas and so
it is the L in LIP. Therefore the correct sum now becomes

```
    C   U   P
    A   N   D
    ?   I   P
  ─────────────
  S   L   I   P
```

So SLIP is a 4-figure square with four different non-zero digits, P is the final digit of a
square, P+D = 10 and P≠5. In particular, if S=1 then P=4 or 6. The only possibilities are
1296   1764   1936   (the next being 2916, too large).

So S=1 and P/D are 4/6 in some order. Also U+N=9 and U/N must be 2/7 in some order.
That means that SLIP can only be 1936, making S=1, L=9, I=3, P=6, D=4, U/N=2/7,
leaving C(>A) as 8 and A=5. Then we must have ?=5 to complete the sum:

```
    8   2/7   6
    5   7/2   4
    5   3     6
  ─────────────
  1   9   3   6
```

But for the three numbers to be in decreasing order they are uniquely determined as
826, 574 and 536.

72 DON'T MISS A SECOND

Answer: 11 and 26th

Let the loss on the wall clock be 's'/'p' where 's' is the number of seconds lost over a period of 'p' seconds, with 'p' being an integer less than 100.

For the wall clock to go, other than in reverse, then 's'/'p' must be less than 1, so 's' < 'p'. The gain on the 24-hour clock must then be 1.025 x s/p.

After 't' seconds the counted number of seconds on the losing clock is t(1 − s/p) and on the gaining clock the counted number of seconds is t(1 + 1.025 x s/p).

Both clocks will show the same time when their number of seconds, counted from their last showing the same time is the same after subtraction of the time taken on their complete 12 or 24 hour cycles. The cycle of the wall clock is 12 hours which is 12 x 60 x 60 = 43,200 secs and the 24-hour clock is twice this at 86,400 secs.

If the number of completed cycles of the 24-hour gaining clock is 'a' and that of the 12-hour losing clock is 'b' then for both clocks to show the same time −

$$t(1 + 1.025 \times s/p) - t(1 - s/p) = 86{,}400a - 43{,}200b \quad \text{which simplifies to}$$

$$2.025 \times s/p \times t = 43{,}200(2a - b)$$

Both clocks display the same time after 'n' days (6 or less) plus 23 hours from their last displaying the same time. So t = 86,400n + 82,800 secs.

$$(2.025 \times s/p)(86{,}400n + 82{,}800) = 43{,}200(2a - b)$$

Let y = (2a - b), for which y must be an integer greater than 0, this then simplifies to

$$\frac{s}{p} = \frac{160}{27} \times \frac{y}{(24n + 23)}$$

s is an integer less than 100 and p is an integer greater than s, but again less than 100. This is only possible if (24n + 23) and 160 have a common factor that is greater than one. The following table shows the values of (24n + 23) for values of 'n' between 0 and 6, and factors for these values.

n	0	1	2	3	4	5	6
(24n + 23)	23	47	71	95	119	143	167
factors	1, 23	1, 47	1, 71	1, 5, 19, 95	1, 7, 17, 119	1, 11, 13, 143	1, 167

It can be seen that a solution only exists for n = 3 for which (24n + 23) = 5 x 19.

s/p = (160/27) x (y/95) = (32/27) x (y/19). Now s<p, so y is 16 or less. The only value of y which allows whole numbers for s and p is y = 9. Then s/p = 32/57.

The wall clock therefore loses 32 seconds every 57 seconds. It needs to be set to the correct hourly time when (3600h + 15x60) x 32 / 57 is an exact multiple of 43,200 seconds.

If the multiple is 'm' then (3600h + 15x60) x 32 / 57 = 43,200m. This simplifies to 8h + 2 = 171m.

On inspection the smallest value of h, and hence the closest time, to satisfy this equation is h = 128 (and m = 6) The time has therefore to be set 128 hours and 15 minutes before kick-off which equates to 5 days 8 hours 15 minutes which is 11:00 on the 26th of the month.

Following a similar method for the 24hr clock gives an equation $41(4h + 1) = 6,840m$ which it can be seen has no solution. The 24-hour clock can therefore never be correctly set on the hour to give the exactly correct time at kick-off.

73   PIN

Answer: 1771

Two 'styles' of number give rise to the required six different arrangements of digits-
a.   3 figures   A, B, C
b.   4 figures   A, A, B, B

The square cannot be 4. It cannot be 9 or 36 as this would lead to divisibility by 3x3. It also cannot be 25, since A+B+C<25 and A+A+B+B is even.
Therefore the sum of the digits must be 16.

a.   Possibilities are   1,7,8   3,4,9   3,5,8   (neglecting those which must lead to divisibility by 2x2 or 5x5)
3,4,9   gives primes
853   is prime
1,7,8   all six arrangements are products of just two primes

b.   Possibilities are   1,7   2,6   3,5
To aid the calculations, note that AABB and ABBA are divisible by 11 and ABAB is divisible by 101.

1,7   **1771 = 7 x 11 x 23. All five rearrangements are the product of just two primes:**
           1177 = 11 x 107
           1717 = 17 x 101
           7117 = 11 x 647
           7171 = 71 x 101
           7711 = 11 x 701
3,5   3355 = 5 x 11 x 61
           3535 = 5 x 7 x 101
           3553 = 11 x 17 x 19
           5335 = 5 x 11 x 97
           5353 = 53 x 101
           5533 = 11 x 503
           no unique case
2,6   2662 = 3x11x11x11 - not allowed

74 THE PLUMBER'S BUCKETS

Answer: 13, 17 and 19 litres

The plumber could not empty the tank by only using one bucket so none of them has a capacity of 10 or 20 litres. The table below shows the amounts that a bucket with capacity (C) in the range 11-19 litres can remove from the tank.

Bucket Capacity (C)	1 x C	2 x C	3 x C	4 x C	5 x C	6 x C	7 x C	8 x C	9 x C
11	11	22	33	44	55	66	77	88	99
12	12	24	36	48	60	72	84	96	
13	13	26	39	52	65	78	91		
14	14	28	42	56	70	84	98		
15	15	30	45	60	75	90			
16	16	32	48	64	80	96			
17	17	34	51	68	85				
18	18	36	54	72	90				
19	19	38	57	76	95				

The dent in the smallest bucket reduced its capacity by 3 litres and this reduced the capacity removed from the tank by 6 litres. This could only happen by filling the smallest bucket twice. Each bucket was used a different number of times so the capacity removed by each bucket must be selected from a different row and a different column of the table, with the capacity removed by the smallest bucket being selected from the second column. There are 10 possible three-bucket combinations as shown in the table below.

Three-Bucket Combinations	Two-Bucket Combinations
2 x 11 + 5 x 12 + 1 x 18 = 22 + 60 + 18 = 100	8 x 11 + 1 x 12 = 88 + 12 = 100
2 x 11 + 1 x 14 + 4 x 16 = 22 + 14 + 64 = 100	6 x 14 + 1 x 16 = 84 + 16 = 100
2 x 11 + 4 x 15 + 1 x 18 = 22 + 60 + 18 = 100	5 x 11 + 3 x 15 = 55 + 45 = 100
2 x 12 + 4 x 15 + 1 x 16 = 24 + 60 + 16 = 100	3 x 12 + 4 x 16 = 36 + 64 = 100
2 x 13 + 1 x 14 + 4 x 15 = 26 + 14 + 60 = 100	5 x 14 + 2 x 15 = 70 + 30 = 100
2 x 13 + 4 x 14 + 1 x 18 = 26 + 56 + 18 = 100	2 x 14 + 4 x 18 = 28 + 72 = 100
2 x 13 + 1 x 17 + 3 x 19 = 26 + 17 + 57 = 100	
2 x 14 + 1 x 15 + 3 x 19 = 28 + 15 + 57 = 100	5 x 14 + 2 x 15 = 70 + 30 = 100
2 x 15 + 1 x 16 + 3 x 18 = 30 + 16 + 54 = 100	4 x 16 + 2 x 18 = 64 + 36 = 100
2 x 15 + 3 x 17 + 1 x 19 = 30 + 51 + 19 = 100	1 x 15 + 5 x 17 = 15 + 85 = 100

The plumber could not empty the tank by only using two buckets but the right-hand column of the table shows that this is possible for nine of the ten three-bucket combinations. The only combination that requires three buckets to empty the tank is the one with bucket capacities of 13, 17 and 19 litres. In this case, it should be possible to empty the tank using the smallest bucket twice, the middle-size bucket once and the largest bucket three times, except that the dent in the smallest bucket causes 6 litres to remain in the tank.

It should also possible to empty the tank using the smallest bucket once, the middle-size bucket four times and the largest bucket once because 1 x 13 + 4 x 17 + 1 x 19 = 13 + 68 + 19 = 100, but this is not valid because the dent in the smallest bucket would cause 3 litres to remain in the tank.

75 COLOURFUL CHARACTERS

Answer: 36 red, 67 yellow, 55 blue and 42 green

The first point is to consider the candidates and there are some short cuts. Starting with green, if a number is to be divisible by single-digit prime, the candidates are 2, 3, 5 and 7.

$3 \times 5 \times 7 = 105$, which is too high and thus we know that 2 must be there and the number of green balls must therefore be even:

$2 \times 3 \times 5 = 30$ so 60 is also possible

$2 \times 3 \times 7 = 42$

$2 \times 5 \times 7 = 70$ too high

so the green candidates are 30, 42 and 60

Now, looking at the yellows, all prime numbers apart from 2 are odd:

31 37 41 43 47 53 59 61 67

We now realise that, with the total ending in zero and, already having one definitely even number and one definitely odd number, the other two must be of opposite polarity if there is to be any chance of the total ending in zero to be correct. So, looking at the red squares, we have

36 49 64

and the palindromes, obviously multiples of 11, we have

33 44 55 66

We now list the candidate totals of these two, add possible green candidates and see if we can find a prime to fit.

Combination	Deduct total from 200	try each of green candidates	yellow
33 and 36	131	out	
33 and 64	103	42	61
		60	43
44 and 49	107	60	47
55 and 36	109	42	67
55 and 64	81	out	
66 and 49	85	42	43

Thus we have the following possibilities:

Red	Yellow	Blue	Green
64	61	33	42
64	43	33	60
49	47	44	60
36	67	55	42
49	43	66	42

So we see that, only if Martha knew that the number of red balls was 36, could she do the whole roll call.

76 GERMOMETRIC MEAN

Answer: 96 72 64 54 48 32 16

The seven data values all differ – decreasing 2-fig. values from Monday to Sunday.

For GM integer the data product must have prime factors to power 7 (or multiple of) For 7 different 2-fig. values only 2, 3, 5, 7 can be involved, because for 11 – only 1x11 to 9x11 valid – no heptad gives product with 7th power for some prime factors needed.

Prime factor	Two-fig. Options from 10, 12, 14, 15, 16, 18, 20, 21, 24, 25, 27, 28, 30, 32, 35*, 36, 40, 42, 45, 48, 49, 50, 54, 56, 60, 63, 64, 70*, 72, 75, 80, 81, 84, 90, 96, 98
7	2.7; 3.7; 2.2.7; [5.7]*; 2.3.7; 7.7; 2.2.2.7; 3.3.7; [2.5.7]*; 2.2.3.7; 2.7.7
5	2.5; 3.5; 2.2.5; 5.5; 2.3.5; 2.2.2.5; 3.3.5; 2.5.5; 2.2.3.5; 3.5.5; 2.2.2.2.5; 2.3.3.5
3	2.2.3; 2.3.3; 2.2.2.3; 3.3.3; 2.2.3.3; 2.2.2.2.3; 2.3.3.3; 2.2.2.3.3; 3.3.3.3; 2.2.2.2.2.3
2	2.2.2.2; 2.2.2.2.2; 2.2.2.2.2.2.2

GM=data val.; 2xGM=data val. (no 3xGM val.); GM/3=data val. (no GM/2 val.), so GM is divisible by 2 and 3 and 30<=GM<=48 – so GM=30 or 36 or 42 or 48

GM	Product of 7 values	GM/3	2xGM	3xGM	GM/2	Contribution from 4 other different data
30	[2^7][3^7][5^7]	10	60	90	15	[2^3][3^5][5^4]
36	[2^14][3^14]	12	72	na	18	[2^7][3^9]
42	[2^7][3^7][7^7]	14	84	na	21	[2^3][3^5][7^4]
48	[2^28][3^7]	16	96	na	24	[2^15][3^5]

None involve 5 and 7 together so data 35=5.7 and 70=2.5.7 – invalid *

GM=30: 4 data contribution excl. 10 30 60 (15 90) is [2^3][3^5][5^4]=[5.5][2.5.5][3^4] [2.2.3] or [5.5][2.5.5][3^3][2.2.3.3] etc.
not unique sol. – invalid

GM=36: 4 data contribution excl. 12 36 72 (18) is [2^7][3^9]=[3^4][3^3][2.2.2.3][2.2.2.2.3] only
Unique data set is [81, 72, 48, **36**, 27, 24, 12], but GM on Thursday – invalid

GM=42: 4 data contribution excl. 14 42 84 (21) is [2^3][3^5][7^4]=[7.7][2.7.7][3^4][2.2.3] or [7.7][2.7.7][3^3][2.2.3.3] etc.
not unique sol. – invalid

GM=48: 4 data contribution excl. 16 48 96 (24) is [2^15][3^5]=[2^6][2^5][2.2.2.3.3] [2.3.3.3] only

Unique data set is [96 72 64 54 48 32 16] and GM on Friday – valid

77   POLL POSITIONS

Answer: A33, B35, C36, D36

All the voter rankings were different, so in the first round no candidate got more than 6 first preferences. After D was eliminated, C was ranked 2nd on 1st preferences because it was B that was eliminated next. If the numbers of first preferences for A, B, C, D was a, b, c, d we necessarily have

$$a \leq 6, \quad a > b > c > d, \quad (c + d) > b$$

The possibilities for (a, b, c, d) are: i] (6, 5, 4, 3) ii] (6, 5, 4, 2) iii] (6, 4, 3, 2) iv] (5, 4, 3, 2)

i] (6, 5, 4, 3) The rankings of voters who put D first (we call these 'D's rankings') had to be DCAB, DCBA and DAXY giving C 6 1st s when D is eliminated, beating B's 5 1st s. When B is eliminated C can get to 9 1st s if B's rankings include BCDA, BCAD, BDCA, but cannot get the stated majority of 10 1st s.

ii] (6, 5, 4, 2)
We give the slips that allow C to beat B and then A and maximise D's points. D can't get to president either way.
ABCD, ABDC, ACBD, ACDB, ADBC, ADCB, BCAD, BCDA, BDAC, BDCA, <u>BXDY</u>, CDAB, CDBA, CADB, CBDA, DCBA, DCAB:
Points with <u>BXDY = BCDA</u>: A43, B 42, C 46, D41   Points with <u>BXDY = BADC</u>: A45, B 42, C 44, D41

iii] (6, 4, 3, 2)
We give the slips that allow C to beat B and then A and maximise D's points. D can't get to president either way.
ABCD, ABDC, ACBD, ACDB, ADBC, ADCB, BCAD, BCDA, BDAC, BDCA, CDAB, CDBA, <u>CXDY</u>, DCBA, DCAB:
Points with <u>CXDY = CBDA</u>: A37, B36, C40, D37   Points with <u>CXDY = CADB</u>: A39, B34, C40, D37

iv] (5, 4, 3, 2)
We give the slips that allow C to beat B and then A and maximise D's points.
<u>AXYD</u>, ABDC, ACDB, ADBC, ADCB, BCAD, BCDA, BDAC, BDCA, CDAB, CDBA, <u>CPDQ</u>, DCBA, DCAB:
Points with <u>ACBD</u> and <u>CADB</u> : A35, B32, C37, D36
Points with <u>ACBD</u> and <u>CBDA</u> : A33, B34, C37, D36
Points with <u>ABCD</u> and <u>CADB</u> : A35, B33, C36, D36 with order D=C>A>B
Points with <u>ABCD</u> and <u>CBDA</u> : A33, B35, C36, D36 with order D=C>B>A giving the solution
The only arrangement of slips that allows the order D=C>B>A has Borda points A33, B35, C36, D36.

Answer: 1027

Denote the square number by n^2, and then the triangular number that is the sum of all integers up to n will equal $\frac{1}{2}n(n+1)$. Therefore the two of them together total $\frac{1}{2}n(3n+1)$, which is what the value of the circular number is required to be. So theoretically a solver might reach the intended solution by exhaustively trying increasing values of n and looking for one that has $\frac{1}{2}n(3n+1)$ satisfying the definition of a circular number in base 1501. However, the Teaser conditions were intended to make the necessary n sufficiently large that solvers would give up on that approach!

Circular numbers are also known as automorphic numbers. If we have a number base q and a circular number x having k digits, then the remainder when x^2 is divided by q^k must simply be x, so that the last k digits of are x^2 precisely the k digits of x itself. Therefore q^k must divide x^2-x exactly. For instance, in base 10, $376^2 = 141376$, and 10^3 divides $141376 - 376$ exactly. Obviously the circular number must have at least one digit (i.e. $k \geq 1$). So, at least, $1501 = 19 \times 79$ has to divide $x^2-x = x(x-1)$ exactly. But x and $x-1$ can't have any common factors. Therefore, since 19 & 79 are primes, there are four possibilities, as follows. (Two of them are the relatively trivial possibilities of the units digit being 0 or 1.)

If 19 divides x and 79 divides $x-1$, then, for some non-negative integer u, $x = 79u + 1$, which must be divisible by 19. Considering the remainder on dividing by 19, $u = 0$ gives a remainder of 1, and each increase of 1 in u adds a remainder of 3. So, in particular, $u = 6$ gives the lowest x that is divisible by 19; and also, adding further multiples of 19 to u will give all other feasible values of x. So the general solution to consider is $x = 79 \times (6+19s) + 1 = 475 + 1501s = 19 \times (25+79s)$. If $s = 0$ then $x = 475$ has only one digit in base 1501, but x can't equal $\frac{1}{2}n(3n+1)$ because $n = 17$ gives 442 (too small) and $n = 18$ gives 495 (too big). The sub-cases with s bigger than 0 will be dealt with later.

If 79 divides x and 19 divides $x-1$, then for some non-negative integer v, $x = 79v$ and 19 must divide $79v - 1$. Considering remainders on dividing by 19, as before, leads to a general solution of $v = 13 + 19t$, giving $x = 79 \times (13+19t) = 1027 + 1501t$. If $t = 0$ then $x = 1027$ is a valid solution for the Teaser because it equals $\frac{1}{2}n(3n+1)$ when $n = 26$. Again, sub-cases with t bigger than 0 are dealt with later.

If 1501 divides x, we know x isn't 0 because it's at least 18; so write $x = b \times 1501^k$ for b being non-divisible by 1501 and k being a positive integer. Then $x^2 = b^2 \times 1501^{2k}$ and the $(k+1)$th digit from the end is zero, but the corresponding digit in x is nonzero, so this cannot work.

If 1501 divides $x-1$, we know x isn't 1 because it's at least 18; so correspondingly write $x = b \times 1501^k + 1$; the last digit of this is 1, preceded by exactly $k-1$ zero digits. Then $x^2 = b^2 \times 1501^{2k} + 2b \times 1501^k + 1$; for this to end in precisely the digits of x we would need $b = 0$, which is a contradiction.

The above is the approach that solvers would use to reach the solution. That analysis covered all cases where x had one digit. However, to establish the uniqueness of the Teaser solution we need to consider s or t greater than 0 in the expressions above and show that there are no other legal solutions.

For x to have three or more digits would need s or t to be at least 1501, which would make x exceed a million, which is prohibited by the Teaser conditions.

All that remains is when x has two digits, and then we need 1501^2 to divide $x(x-1)$ exactly. Following the analysis above, there are two cases, as follows.

If 19^2 divides $x = 19 \times (25+79s)$ and 79^2 divides $x-1 = 474 + 1501s = 79(6+19s)$, then: 19 divides $25+79s$, for which the general solution can be found to be $s = 17+19y$; and 79 divides $6+19s$, for which the general solution can be found to be $s = 8+79z$. (The constant parts of these results can be found by considering remainders on dividing by 19 and 79 respectively, as done earlier.) Equating the two expressions for s then gives the general solution $y = 12+79w$, $z = 3+19w$ (again, considering remainders), which leads to $s = 245+1501w$ and thence to $x = 475 + 1501 \times 245 + 1501^2 w$, for which 368220 is the only value below a million (occurring when $w=0$). This can't equal $\frac{1}{2}n(3n+1)$ for any integer n because $n=495$ gives 367785 and $n=496$ gives 369272.

If 79^2 divides $x = 79 \times (13+19t)$ and 19^2 divides $x-1 = 1026 + 1501t = 19 \times (54+79t)$, then a similar method looks for 79 to divide $13+19t$, giving $t = 70+79z$, and for 19 to divide $54+79t$, giving $t = 1 + 19y$, which together give $y = 66+79w$ & $z = 15+19w$ and thence $t = 1255+1501w$ and $x = 1027 + 1501 \times 1255 + 1501^2 w$. 1884782 is the smallest value of this (when $w=0$) and is already over a million.

79 PRODUCT DATES

Answer: 274

To find the shortest interval between consecutive product dates, we note that two such product dates are 2 January 2002 and 1 February 2002, with an interval of 29 days between them. In any one month, there cannot be more than one product date, so to find the shortest interval between consecutive dates we need to consider pairs of product dates falling in consecutive months, with an interval of less than 29 days between them.

First let us look to see whether there is a pair of consecutive product dates with the first on Nth December one year, and the second on Mth January of the next year. Then the first of these years is 2000+12N and the second is 2001+12N, so 1+12N= 1*M. Hence N = 1 or 2, giving M=13 or 25 respectively, both of which produce intervals of more than 30 days between the consecutive product dates. So we consider now a pair of consecutive product dates in consecutive months of the same calendar year. Suppose that pair is the Nth day of the Xth month, and the Mth day of the X+1th month, where X is 1, 2, 3, ... 11. Then N*X = M*(X+1), so (M-N)*X = -M. The interval V between the pair of dates is M + K - N, where K is the number of days in month X, i.e. 28, 29, 30 or 31.

Therefore V = K − M/X.

If X=1, K = 31 and N=2*M, and the largest possible value of N, which must be even, is 30, so M=15; then V= 31-15 = 16.

If X=2, K=28 or 29 and N=3*M/2 so the largest possible value of N, which must be divisible by 3, is 27, then M=18, and V=19 or 20.

If X>2, K = 30 or 31 and V > 30 − 31/3 > 19.

Hence S, the shortest interval between consecutive product dates, is 16 days, between 30 January 2030 and 15 February 2030.

To find L, we note that 28 December 2336, 29 December 2348, 30 December 2360 and 31 December 2372 are all product dates. There is no other product date falling in the intervals between them, nor in the rest of the millennium after 31 December 2372. The intervals between these four dates are all equal to 12 years and a day, with 3 of the intervening years being leap years. So those intervals all equal 12*365 + 3 + 1, which is 4384 days. Prior to 28 December 2336, there are product dates in other calendar months falling between consecutive product dates in December. So L= 4384.

Now 4384 = 16* 274, so the multiple is 274.

80 HIDDEN POWERS

Answer: 198752346

The powers less than 2022 that have no zeros or repeated digits are:

Squares: 1 4 9 16 25 36 49 64 81 169 196 256 289 324 361 529 576 625 729 784 841 961 1296 1369 1764 1849 1936

Other cubes: 8 27 125 216 512 1728

Other powers: 32 128 243

Note that, a single-figure number can be produced in the game if and only if its digit is in one of the three middle positions. So none of 1, 4, 8 or 9 is in a middle position.

The lowest power possible (between 1066 and 2022) must be one of
1296 1369 1728 1764 1849 1936.

Lowest power = 1296?
To produce 1296 but not the square 196 there must be at least six cards between the 1 & 9:

1							9	6

To avoid 36 and 49 the 3 and 4 must be 2nd and 7th respectively. But then 324 is possible.

Lowest power = 1369?
Likewise to avoid 169 we must have

1	4					3	9	6

But then 27 or 729 will be possible.

Lowest power = 1728?
Likewise to avoid 128 the 1, 2, 8 must be 1st, 8th and 9th. Then to avoid 169 and 196 we must have we must have

1	6/9	9/6					2	8

But then 64 is possible.

Lowest power = 1764?
To avoid the 16 we must have

1							6	4

But then 196 is possible.

Lowest power = 1849?
To avoid 49, 169 and 196 we must have

1						4	6/9	9/6

But then 27 or 729 is possible.

So the lowest power = 1936
As before, to avoid 196 there must be at least six spaces between the 9 and the 6.
We place the 3 to avoid 36, and the 4 and 8 cannot go in the middle three spaces.
So we have one of:

1	9	4				3/8	8/3	6
1	9	8				3/4	4/3	6

In the top row cases 128 will be possible and in the lower row cases the 3 must come before the 4 (to avoid 243). Finally, the 7 must come before the 5 (to avoid 576) and the 5 must come before the 2 (to avoid 25) leaving:

1	9	8	7	5	2	3	4	6

It is easy to check that none of the powers less than 1936 listed at the top of the page can be produced.

Answer: BD 51 HLX

Let N_1, N_2, N_3, N_4 and N_5 be the positions of the five letters in the alphabet in increasing order and let $S = 1/N_1 + 1/N_2 + 1/N_3 + 1/N_4 + 1/N_5$ (Eqn 1). From the teaser statement the N_i's are all different, at most 26 and by the no-vowel condition not 1, 5, 9, 15 or 21. We are seeking values of the N_i for which $S = 1$.

Step 1. With the assumption that $S = 1$, we first rule out the possibility that $N_1 \geq 3$. Suppose that $N_1 \geq 3$. Since $1/4 + 1/5 + 1/6 + 1/7 + 1/8 < 0.89$, we see that $N_1 = 3$. Now suppose that $N_2 \neq 4$. As the vowel E is the 5th letter of the alphabet, $N_2 \neq 5$, and because the maximum value of S is then $1/3 + 1/6 + 1/7 + 1/8 + 1/9 < 1$, it follows that $N_2 = 4$ and $S = 1/3 + 1/4 + 1/N_3 + 1/N_4 + 1/N_5$, and we can assume that $N_3 \geq 6$. Since $1/3 + 1/4 + 1/7 + 1/8 + 1/9 = 0.96$, we deduce that $N_3 = 6$. Similarly, from the inequality $1/3 + 1/4 + 1/6 + 1/8 + 1/9 < 1$, it then follows that that $N_4 = 7$; therefore $S = 1/3 + 1/4 + 1/6 + 1/7 + 1/N_5 = 1$.
For $N_5 = 8$ or 9 the sum S is then strictly greater than 1 and for $N_5 \geq 10$ it is strictly less than 1. This rules out the case $N_3 \geq 6$, and so excludes the possibility that $N_1 \geq 3$. **Therefore $N_1 = 2$.**

Step 2. Next suppose that $N_2 = 3$ and show no solutions exist for $S = 1$. Set $T = 1 - (1/2 + 1/3 + 1/N_3) = 1/6 - 1/N_3 = 1/N_4 + 1/N_5$. Because $1/18 + 1/19 + 1/20 < 3/18 = 1/6$, we have $7 \leq N_3 \leq 17$.

N_3	7	8	10	11	12	13	14	16	17
T	1/42	1/24	1/15	5/66	1/12	7/78	2/21	5/48	11/102

Set $a = N_4$ and $b = N_5$ and conclude that $T = 1/a + 1/b$ with $a < b$. Suppose $T = 1/N$ for some integer N. Then $ab = N(a + b) > 2Na$. Therefore $b > 2N$, and since $b \leq 26$, it follows that $N < 12$. This rules out $N_3 = 7$, 8 and 10.
If **$N_3 = 11$**, $5ab = 66(a + b) > 132a$, and so $b > 132/5 > 26$.
If **$N_3 = 12$**, $ab = 12(a + b) > 24a$, and so $b = 25$ or 26. Substituting $b = 25$ gives $12a = 25a - 12 \times 25$ and so a is divisible by 25. Since a is also divisible by 3, we have $a \geq 75$. If $b = 26$, a is divisible by 13 and 3 and is ≥ 39.
If **$N_3 = 13$**, $7ab = (6 \times 13)(a + b) > 156a$, whence $b > 156/7 > 22$ and $23 \leq b \leq 26$. If $b = 23$ or 25, a is divisible by, and is therefore not less than, 6×13. If $b = 24$, a is again divisible by 13, than $a + b = 13 + 24 = 37$ or $(26 + 24) = 50$ and neither 37 or 50 divide 7ab. Finally, if $b = 26$, $14a = 6(a + 26)$; hence $8a = 6 \times 29$, and a is not an integer.
If **$N_3 = 14$**, $2ab = 21(a + b)$, and as before, we get $22 \leq b \leq 26$. Since these values of b, apart from 24, are coprime with 21, a is a multiple of 21 and therefore equal to 21. Then $2b = 21 + b$, and $b = 21$ is out of range. If $b = 24$, then $16a = 7(a + 24)$ and $a = 7 \times 24/9$, which is not an integer.
If **$N_3 = 16$**, $5ab = 48(a + b)$ and $b \geq 20$. For $b = 20$, $a = 960/52$, which is not an integer. For $b = 21$ we deduce that $a = 960/57$, again not an integer. For $b = 22, 23, 25, 26$, a is a multiple of 24 and therefore equal to 24, Since $(a + b)$ then has the values, 46, 47, 49, 50, none of which divide 5ab, there remains the possibility that $b = 24$. But with $b = 24$ we get $a = 16$ and N_3 and N_4 are equal, against hypothesis. This settles $N_3 = 16$.
If **$N_3 = 17$**, then $N_4 \geq 18$. Since $1/17 + 1/18 + 1/19 > 1/6$ and $1/17 + 1/18 + 1/20 < 1/6$, this case cannot arise. Therefore **$N_2 \geq 4$.**

Step 3. We next try $N_2 > 4$. Since no N_i can equal to 5 or 9 and since $1/2 + 1/6 + 1/8 + 1/10 + 1/11 < 0.99$, we conclude that $N_2 = 6$ and $N_3 = 7$. Set $a = N_4$ and $b = N_5$ so we can write $4/21 = 1 - (1/2 + 1/6 + 1/7) = 1/a + 1/b$ with $a < b$. Hence $(a + b)/ab = 4/21$ (Eqn b) and therefore $b = 21a/(4a - 21) > a$, which implies that $a < 10.5$. The only possibilities are $a = 8$ or 10 and in neither case is b an integer when these values of a are substituted in Eqn b. Therefore **$N_2 = 4$.**

Step 4. Finally we show the only possibility is **$S = 1/2 + 1/4 + 1/8 + 1/12 + 1/24 = 1$** (Eqn c). Consider the possibility that $N_3 = 6$. Then $1/12 = 1 - (1/2 + 1/4 + 1/6) = 1/a + 1/b$ with $a < b$, as before $b = 12a/(a - 12) > a$, whence $a < 24$. The inequality $b < 27$ yields $a > 22$. The only option is $a = 23$ and then $b = (12 \times 23)/11$, which is not an integer. A similar argument excludes the possibilities that $N_3 = 7$ and 11. Since $1/2 + 1/4 + 1/12 + 1/13 + 1/14 < 0.99$, it follows that $N_3 = 8$ or 10. Suppose that $N_3 = 8$. As before we set $1/8 = 1 - (1/2 + 1/4 + 1/8) = 1/a + 1/b$ with $a < b$. It follows that $ab = 8(a + b)$ and hence that $b = 8a/(a - 8) > a$ and we conclude that $a < 16$. Moreover the inequality $b \le 26$ implies that $a \ge 12$. Setting $a = 12$ gives $b = 24$, and since the values $a = 13, 14, 15$ fail to make b an integer, Eqn c is the unique solution to Eqn a in the case $N_3 = 8$. Using the corresponding version of Eqn b once more as above, we see that the case $N_3 = 10$ yields one further solution to Eqn a, namely $1 = 1/2 + 1/4 + 1/10 + 1/12 + 1/15$. but this is again ruled out since 15 corresponds to the vowel O. This completes Step 4 and shows that **the registration plate reads BD 51 HLX.** (Thus the vehicle was therefore first registered between 1st September 2001 and 31st March 2002.)

Answer: 105

Let the number of questions set be 'q' and the number of marks given for a correct answer be 'm'.

The number of different mark totals or scores which can be obtained is equal to q (the negative totals if no correct answers are given with various numbers of incorrect answers) <u>plus</u> 1 (zero total marks) <u>plus</u> q x m (maximum number of positive totals) <u>less</u> number of positive totals which cannot be attained.   (i)

The formula for the number of unattainable positive totals can best be had by considering the following examples consisting of six questions. There are two situations, where q ≥ m and where q ≤ m, with m being 4 and 8 in the examples.

<table>
<tr><td colspan="2" align="center">q ≥ m</td><td colspan="2" align="center">q ≤ m</td></tr>
<tr><td colspan="2" align="center">q = 6   m = 4</td><td colspan="2" align="center">q = 6   m = 8</td></tr>
<tr><td align="center">No. correct answers</td><td align="center">Number of unattainable totals</td><td align="center">No. correct answers</td><td align="center">Number of unattainable totals</td></tr>
<tr><td align="center">1</td><td align="center">4 - 6 =   *</td><td align="center">1</td><td align="center">8 - 6 =   2</td></tr>
<tr><td align="center">2</td><td align="center">4 - 5 =   *</td><td align="center">2</td><td align="center">8 - 5 =   3</td></tr>
<tr><td align="center">3</td><td align="center">4 - 4 =   0</td><td align="center">3</td><td align="center">8 - 4 =   4</td></tr>
<tr><td align="center">4</td><td align="center">4 - 3 =   1</td><td align="center">4</td><td align="center">8 - 3 =   5</td></tr>
<tr><td align="center">5</td><td align="center">4 - 2 =   2</td><td align="center">5</td><td align="center">8 - 2 =   6</td></tr>
<tr><td align="center">6</td><td align="center">4 - 1 =   3</td><td align="center">6</td><td align="center">8 - 1 =   7</td></tr>
<tr><td align="center">total unattainable</td><td align="center">6</td><td align="center">total unattainable</td><td align="center">27</td></tr>
</table>

* cannot be negative.

These are both arithmetical progressions (AP) whose sum is ½(first + last terms) x (number of terms).

AP Sum = ½(0 + 3) x 4

Generalised AP Sum $= \frac{1}{2}(0 + m\text{-}1) \times m$

$\qquad = m(m\text{-}1)/2$

When q = m both AP sums are the same.

No. Totals = $q + 1 + qm - m(m\text{-}1)/2 = 100$

rearranged: $q = (m^2 - m + 198)/2(m + 1)$ (ii)

See below

AP Sum = ½(2 + 7) x 6

Generalised AP Sum $= \frac{1}{2}(m\text{-}q + m\text{-}1) \times q$

$\qquad = q(2m - q - 1)/2$

Substituting in equation (i) gives –

No. Totals = $q+1+qm-q(2m-q-1)/2 = 100$

rearranged: $q^2 + 3q - 198 = 0$

The quadratic formula shows that this doesn't give an integer solution for q therefore m cannot be more than q.

This table shows the calculated value of q, from equation (ii), for various values of m. Values of m above 12 result in q being less than m, so can be ignored.

m	q	m	q	m	q
1	49.5	5	18.2	9	13.5
2	33.3	6	16.3	10	13.1
3	25.5	7	15	11	12.8
4	21	8	14.1	12	12.7

Integer values of q are only obtained for values of m of 4 and 7. Therefore 15 questions could be asked with 7 marks for a correct answer or 21 questions and 4 marks for a correct answer. The highest number of marks that a student can get is 15 x 7 = 105.

83 BANK ROBBERY

Answer: Knife, motor bike

The laborious method of solution involves trial and error, exploring 2 facts at a time through those given. A smarter way of solving the puzzle is to realise that, since everyone got 2 facts correct, there were 10 facts right altogether.

Drawing up a table of observations we have:

Witness	Height	Hair colour	Eye colour	Weapon	Escape method
1	Short	Fair	Brown	Cricket bat	Motor bike
2	Tall	Fair	Grey	Gun	Car
3	Tall	Dark	Brown	Crowbar	Motor bike
4	Short	Ginger	Blue	Knife	Car
5	Tall	Dark	Blue	Stick	Push bike

As we require 10 correct observations, 0, 2, or 3 had correct height
0, 1, or 2 had correct hair colour
0, 1, or 2 had correct eye colour
0 or 1 had correct weapon
0, 1 or 2 knew the method of escape correctly.
To have 10 correct altogether, we must have 3 heights, 2 hair colours, 2 eye colours, 1 weapon and 2 escape methods correct.
This narrows the field to: a tall person with fair or dark hair, brown or blue eyes and using a car or motor bike.
But to get 2 facts right per witness, this narrows further to a tall man with fair hair and blue eyes travelling by motor bike. For witness 4 to have 2 details right, he must have been carrying a knife.

Thus, the robber carried a knife and escaped by motor bike.

84 A SIX-PIPE PROBLEM

Answer: 420cm

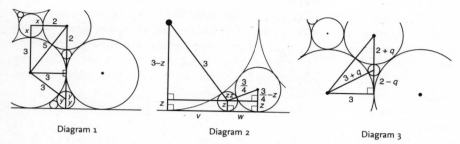

Diagram 1   Diagram 2   Diagram 3

Let the radii of the As and Bs be 3 and 2 units (not cm). In Diagram 1, by symmetry, the quadrilateral has vertical and horizontal sides and is a rectangle. Then the upper triangles are 3-4-5 and the radius of type C is $x = 1$.

Let the type D have radius y. In the lower right-angled triangle in Diagram 1

$(3 + y)^2 = 3^2 + (3 - y)^2$ which is $6y = 9 - 6y$ and $y = 3/4$

In the right-angled triangles in Diagram 2, $v^2 = (3 + z)^2 - (3 - z)^2$ and $w^2 = (3/4 + z)^2 - (3/4 - z)^2$.

So $v^2 = 12z$ and $w^2 = 3z$ so $v + w = 3\sqrt{3}(\sqrt{z}) = 3$ and the radius of type E is $z = 1/3$.

In the smaller right-angled triangle in Diagram 3

$(3 + q)^2 = 3^2 + (2 - q)^2$ so $10q = 4$ and the radius of type F is $q = 2/5$.

So the six pipe radii (in units, not cm) are 3, 2, 1, 3/4, 2/5, 1/3

To get these sizes to minimal whole-number diameters in cm we must multiply by 60 to give 180, 120, 60, 45, 24, 20. The final stack height consists of two type-A and one type-C pipes. If the diameters were multiples of the above numbers, the final height would be more than 5m, so only the given diameters are possible and **the final height of the stack is 180 + 60 + 180 = 420cm.**

85 LAWN ORDER

Answer: 176

Without loss of generality ('wlog'), think of the edges being aligned with an x-y coordinate grid. Then as someone proceeded around the perimeter, they would move to either Right (increasing y), Up (increasing y), Left (decreasing x) or Down (decreasing y). Because the edges must alternate between Left-or-Right and Up-or-Down around the perimeter, the number of Left plus Right, and Up plus Down, edges must be equal – therefore they must both equal 5.

In order to form a circuit, the total lengths of Right and Left edges must be equal, and similarly for Up and Down.

Let the 16-foot edge be one that moves one to the Right, wlog. What could the other four Right or Left edges be? If 16 were the only Right edge, then there would have to be four Left edges, all of length 4; but then the remaining edges couldn't be split into two sets with sums equal to each other.

If 16 and one other edge are the only Right ones, then the only possibility is for 16 & 5 being Right, three 7s being Left, 8 & 4 being (without loss of generality) Up, and three 4s being Down. If 16 were one of three Right edges, then no two remaining lengths could be Left edges with an equal total. So there is only one assignment of lengths to directions, and it remains to arrange these lengths into simple legal circuits with no intermediate touching or crossing.

With the 'horizontal' edges being R16, R5, L7, L7 & L7, how could the other edges U8, U4, D4, D4 & D4 fit? Firstly, consider the order of horizontal edges as shown. (For the time being, ignore the lower case letters in square brackets in the following notation – they'll be explained later.)

R16 U8 R5[b] U4 L7 D4 L7 D4 L7 D4 is possible – see pattern A.
R16 U8 R5 D4 L7 etc. fails because that last edge crosses another.
R16 U4 R5[a] U8 L7 D4 L7 D4 L7 D4 is possible – see pattern B.
R16 D4[f] R5 U8 L7 U4 L7 D4 L7 D4 is possible – see pattern C.
R16 D4 R5 U8 L7 D4 etc. has a crossing or intersection caused by that last edge, and so can't form a simple circuit. Ditto R16 U4 R5 D4 L7 etc. Ditto R16 D4 R5 U4 L7 etc. Ditto R16 D4 R5 D4 L7 U8 etc. Ditto R16 U4 R5 D4 L7 etc. Ditto R16 D4 R5 U4 L7 etc. Ditto R16 D4 R5 D4 L7 U4 L7 U8 etc.
R16 D4 R5 D4[g] L7 D4 L7 U8 L7[f] U4 is possible – see pattern D.
R16 D4 R5 D4 L7 U4 L7[f] D4 L7 U8 is possible – see pattern E.
R16 D4[c] R5 D4[h] L7 D4 L7[e] U4 L7[d] U8 is possible – see pattern F.

This exhausts all 'vertical' arrangements here; wlog, the only remaining order of 'horizontal' edges is R16 L7 R5 L7 L7, and all the possibilities with this are as follows.
R16 U8 L7 U4 R5 D4 etc. has a crossing. Ditto R16 U8 L7 D4 R5 U4 etc. Ditto R16 U8 L7 D4 R5 D4 etc. Ditto R16 U4 L7 U8 R5 D4 L7 etc. Ditto R16 D4 L7 U8 etc. Ditto R16 U4 L7 D4 etc. Ditto R16 D4 L7 U4 etc. Ditto R16 D4 L7 D4 R5 U8 etc. Ditto R16 U4 L7 D4 etc. Ditto R16 D4 L7 U4 etc. Ditto R16 D4 L7 D4 R5 U4 etc.
R16 D4 L7 D4[d] R5 D4 L7 U8 L7[h] U4 is possible – see pattern G.
R16 D4 L7 D4[f] R5 D4 L7 U4 L7[g] U8 is possible – see pattern H.

So there are eight possible basic patterns, plus rotations and reflections. The answer to the Teaser will come from the pattern that can be amended in four ways, as suggested by the neighbour, to form other legal patterns. These are indicated in the listing above: [x] after an edge description means that pattern x is obtained by rotating through 180° the linkage formed by that edge and its preceding and following edge. There are three reasons why this sort of amendment might not be 'legal' for the Teaser: if the preceding and following edges are the same length and in the same direction, so that the suggested amendment wouldn't change anything; if the amendment would give the same shape, just rotated/reflected; or if the amendment would give a perimeter that touches or crosses itself at an intermediate point.

Note that the amendment of F to produce C actually produces a shape that is a 180° rotation of the C pattern that arises from the listing shown above; and vice versa.

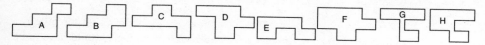

The only pattern that has at least four legal amendments is F, and this has an area of 16×4 + 21×4 + 7×4 = 176 square feet.

86 THE BEARINGS' TRAIT

Answer: 159, 267 and 348°

Three 3-digit bearings (none prime) use all numerals 1-9, so no bearing includes '0'. Thus the bearings are in the E-S [090-180; must be 1--], S-W [180-270; must be 2--] and the W-N [270-360; must be 3--] quadrants.

W-N quad. bearing from: 345, 346, 348, 354, 356, 357, 358

S-W quad. bearing from: 245, 246, 247, 248, 249, 254, 256, 258, 259, 264, 265, 267, 268

E-S quad. bearing from: 145, 146, 147, 148, 154, 156, 158, 159, 164, 165, 168, 169, 174, 175, 176, 178

Tabulate W-N bearing options v. S-W valid options v. E-S options

! - ambiguous   * - unambiguous

345		267		268							
		invalid		invalid							
346	*	258		**259**	!						
		invalid		**178**	!						
348	!	256		**259**	!	265		**267**	*		
		invalid		**176**	!	invalid		**159**	*		
354		267		268							
		invalid		invalid							
356	*	247		248		**249**	!				
		invalid		invalid		**178**	!				
357	!	246		**248**	*	**249**	!	264		268	
		invalid		**169**	!	**168**	*	invalid		invalid	
358	!	246		**247**	*	**249**	!	264		267	
		invalid		**169**	!	**176**	!	invalid		invalid	

Only triad with two * and one ! is **159*, 267*, 348!**

So random choice of any one from these more likely to be certain than not.

Answer: 5, 3, 6, 12 and 7

Let the five numbers of sweets be A, B, C, D and E (since the ten totals give no repeats, it follows that these five are all different).

The lowest pair A+B=8. So the two lowest numbers (A and B in some order) are one of:
1 & 7   or   2 & 6   or   3 & 5
In particular, the next number of sweets up is at least 6.

The second highest pair, C+D=18 and so the top three numbers (C, E, D or D, E, C) are one of

6 E 12   (E=7/8/9/10/11)
7 E 11   (E=8/9/10)
8 E 10   (E=9)

Fitting the bottom pair and top three together gives the following possibilities for A-E in some order:

1 7 8 9 10
2 6 7 E 11   (E=8/9/10)
2 6 8 9 10
3 5 6 E 12   (E=7/8/9/10/11)
3 5 7 E 11
3 5 8 9 10   (E=8/9/10)

But the ten totals of the pairs are all different, so we are left with the following sets for A-E:

A-E in some order	The ten totals	Total of sweets
2 6 8 9 10	8 10 11 12 14 15 16 17 18 19	35
3 5 6 7 12	8 9 10 11 12 13 15 17 18 19	33
3 5 7 8 11	8 10 11 12 13 14 15 16 18 19	34
3 5 7 10 11	8 10 12 13 14 15 16 17 18 21	36

We are told that there is one number between 8 and 18 which identifies just one total number of sweets; i.e. just one row. The only such number is 9. So the numbers of sweets are
{A, B ,C, D, E} = {3, 5, 6, 7, 12}.

But we are then told that the 9 is B+C. Therefore
A+B=8, B+C=9, C+D=18.

This uniquely determines
A=5, B=3, C=6, D=12, E=7.

Answer: Ann, Ben, Dave, Celia, Ben

Table 1 below shows the options for the duties after applying the restrictions for prime number dates, non-primes, Mondays and Wednesdays.

Table 1

Date	1	2	3	4	5	8	9	10	11	12	15	16	17	18	19	22	23	24	25	26	Allocation
	M	T	W	T	F	M	T	W	T	F	M	T	W	T	F	M	T	W	T	F	
Ann	A			A		A	A	A		A	A	A		A		A		A	A	A	4
Ben		B			B				B						B		B				4
Celia		C	C	C	C		C	C	C	C		C	C	C			C	C	C	C	6
Dave		D	D		D				D				D		D		D				2
Ed	E			E		E	E			E	E	E		E		E			E	E	4

From this we can deduce that Ben must cover the 2nd, 5th and 11th as he cannot do both duties on the 19th and 23rd. So removing the other candidates for 2nd, 5th and 11th gives....

Table 2

Date	1	2	3	4	5	8	9	10	11	12	15	16	17	18	19	22	23	24	25	26	Allocation
	M	T	W	T	F	M	T	W	T	F	M	T	W	T	F	M	T	W	T	F	
Ann	A			A		A	A	A		A	A	A		A		A		A	A	A	4
Ben		B			B				B						B		B				4
Celia			C	C			C	C		C		C	C	C			C	C	C	C	6
Dave			D										D		D		D				2
Ed	E			E		E	E			E	E	E		E		E			E	E	4

Celia's first duty must be on the 3rd or 4th and her second and third duties have to be on the 9th and 12th in order to accommodate her six duties. So Ann must cover the 10th and Ed must cover the 8th.

Table 3

Date	1	2	3	4	5	8	9	10	11	12	15	16	17	18	19	22	23	24	25	26	Allocation
	M	T	W	T	F	M	T	W	T	F	M	T	W	T	F	M	T	W	T	F	
Ann	A			A				A			A	A		A		A		A	A	A	4
Ben		B			B				B						B		B				4
Celia			C	C			C			C		C	C				C	C	C	C	6
Dave			D										D		D		D				2
Ed	E					E					E	E		E		E			E	E	4

It can now be deduced that Celia's fifth and sixth duties must be on the 23rd and 26th which rapidly leads to the solution below since Ben must cover the 19th and Dave must cover the 3rd and the 17th.

Date	1	2	3	4	5	8	9	10	11	12	15	16	17	18	19	22	23	24	25	26	Allocation
	M	T	W	T	F	M	T	W	T	F	M	T	W	T	F	M	T	W	T	F	
Ann	A							A			A	A						A			4
Ben		B			B				B						B						4
Celia				C		C				C				C			C			C	6
Dave			D										D								2
Ed							E				E	E				E			E		4

Ann and Ed each do one of the duties on the 15th and 16th. Both are valid selections and the order does not alter the solution.

Answer: 19 tiers; 12 and 5 years.

First, we derive a formula for the number of bonbons in a regular tetrahedral tower having n tiers. In the base of that tower, the number of bonbons is $1 + 2 + ... + n$, which is $n(n+1)/2$. So the number of bonbons in the whole tower is the sum from 1 to n of the expression $x(x+1)/2$, which is half the sum from 1 to n of the expression $x*x + x$, and therefore equals

$\{ n(n+1)(2n+1)/6 + n(n+1)/2\}/2$
$= n(n+1)\{(2n+1)/3 + 1\}/4$
$= n(n+1)(2n+4)/12$
$= n(n+1)(n+2)/6$,

which we denote as $T(n)$.

Using this formula, we are now able to calculate the values of $T(n)$ for $n = 5$ upwards until we reach $T(n) > 2000$. Those values are, successively,
35, 56, 84, 120, 165, 220, 286, 364, 455, 560, 680, 816, 969, 1140, 1330, 1540, and 1771.

The sum of the first eight entries in this list, i.e. $S = 35 + 56 + 84 + 120 + 165 + 220 + 286 + 364$, equals 1330, a later entry in the list, which means that we have one possible arrangement with 8 grandchildren. But are there any possible arrangements with more than that number of children?

The smallest total number of bonbons with 9 children is $1330 + 455$, which equals 1785, greater than any value of $T(n)$ less than 2000. So no solution is possible with 9 children. The same conclusion applies to any arrangement with a higher number of children.

There remains the question as to whether the solution with 8 children is unique. If such an alternative solution is possible, then the sum S would increase to either 1540 or 1771, increases of 210 and 441, respectively. The effect of substitutions of any one of the first eight entries by any one of the later entries is shown in the following table:

The effect of substituting any one of	35	56	84	120	165	220	286	364 by
455 would be	+420	+399	+371	+335	+290	+235	+169	+91
560	+525	+504	+476	+440	+395	+340	+274	+196
680	+645	+624	+596	+560	+515	+460	+394	+316
816	+781	+760	+732	+696	+651	+596	+530	+452

while any substitution by 969 would exceed +441.

None of the entries in this table equals 210 or 441, nor does any combination of the entries. So the arrangement with 8 children given above is unique.

For that arrangement, the total number of bonbons in my original tower was 1330 in 19 tiers, and the oldest and youngest grandchildren have ages 12 and 5 years respectively.

Answer: 171

To receive a score the puck must drop into a corner pocket. From this we know that the puck must have travelled integer multiples of the side length in both directions. Using a grid of table squares we can un-box the puck's path to a straight line. For example, if the puck travelled 3m up-down and 4m left-right, then we know it travelled 5m in total, as shown in Figure 1.

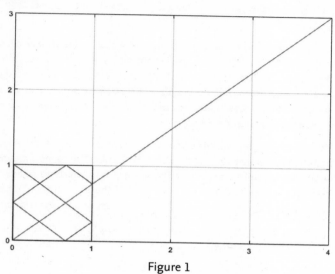

Figure 1

If we say the puck travels x lengths left-right, and y lengths up-down, then its total travel is z, where $z^2 = x^2 + y^2$. We also know that x and y must be co-prime, otherwise the puck would drop into a pocket earlier on its path. Given this we observe that the puck bounces x-1 times left-right, and y-1 times up-down, giving, t = x + y - 2, tokens.

If the puck travels 1m farther than its left-right distance, we can write, $x^2+y^2=(x+1)^2$. Therefore, $y^2 = 2x + 1$. Solutions of this form are generated by, y = 2k+1, x = 2k(k+1), with k = 1,2,3...

If the puck travels 2m farther than its left-right distance, we can write, $x^2+y^2=(x+2)^2$. Therefore, $y^2 = 4(x + 1)$. Solutions of this form are generated by, y = 4k, x = 4k²-1, with k = 1,2,3...[N.B. y = 2k does not satisfy the requirement that x,y are co-prime for odd k].

By applying the above, and searching for solutions where t<1000, it can be found that Claire's cube can only be 125: (x,y,z,t) = (112,15,113,125).

For David's triangle, as Claire won the game, there are two possibilities with t<125: (x,y,z,t) = (15,8,17,21) or (35,12,37,45).

Therefore, after picking up the spare token, they have a total of either 147 or 171 tokens. However, 147 cannot be expressed as the sum of a cube and a square, whilst $171 = 3^3 + 12^2$.

Answer: 1804

There are eight possible squares; only four of them (underlined) do not have repeating digits.

121   144   225   <u>256</u>   <u>324</u>   <u>361</u>   441   <u>625</u>

If a die displays N on one face then its adjacent face cannot be N or 7-N. This restriction means there is only one valid order for the four squares on the vertical faces, namely 256, 324, 625, 361.

Dice can be either left-handed (in which the numbers 1, 2 and 3 appear clockwise around a corner) or right-handed (1,2,3 appear anticlockwise). This gives a choice of two numbers for the corners on the top face, but we must remember that the dice are identical. For left-handed dice we have 256(5), 324(5), 625(1), 361(4) and for right-handed dice we have 256(2), 324(2), 625(6), 361(3), where the number in brackets is the adjacent corner number on the top face.

For right-handed dice, the only possible square on the top face that includes two corners is 256, but the adjacent vertical face will show 625. This is not allowed as the 2 on the top face will be adjacent to the 5 on the vertical face, so there can be no more than two squares on the top face, and we are told that there are three. Therefore, the dice must be left-handed. There are four possible orientations of the top face, as shown.

(a)
```
      5   2   6
  3 | 1       5 | 4
  6 | 4   4   1 | 2
  1 | 4       5 | 3
      2   5   6
```

(b)
```
      1   6   3
  2 | 4   4   1 | 5
  5 |     4     | 2
  6 | 5   1   5 | 6
      3   2   4
```

(c)
```
      6   5   2
  3 | 5       4 | 1
  2 |     4   4 | 6
  4 | 5       1 | 3
      6   2   5
```

(d)
```
      4   2   3
  6 | 5       5 | 6
  2 |         5 | 5
  5 | 1   4   4 | 2
      3   6   1
```

(a) The remaining square (top-bottom) is 144 or 441, with totals <u>1804</u> or 2101
(b) The remaining square (left-right) is 144 or 441, with totals 2101 or 2398
(c) The remaining squares (left-right & top-bottom) are 144&144, 144&441, 324&121 or 324&324, with totals 2299, 2446, 2596 or 2699
(d) The remaining squares (left-right & top-bottom) are 144&144, 121&324, 441&144 or 324&324, with totals 2002, 2149, 2299 or 2402

The only possible total less than 2000 is 1804.

Answer: £1.40

Let the number of times that coin in position 'n' is turned over be 't'. At the end of the game if 't' is even then coin 'n' remains heads up and if odd then it ends tails up.

As the start is to turn every second coin over and not to turn every coin over then 't' is one less than the number of divisors of 'n'. If $n = P_1^a \times P_2^b \times P_3^c \cdot \cdot \cdot \cdot \cdot \cdot$ where Ps are prime numbers, then
$t = \{(a + 1)(b + 1)(c + 1) \cdot \cdot \cdot \cdot \cdot \cdot \cdot\} - 1$.
If any of a, b, c, etc is odd then the bracket and the product of the brackets must be even and 't' must be odd. So for any coins to remain heads up all the powers ie. a, b, c etc must be even, and so 'n' must be a perfect square.

The number of heads must therefore be equal to the number of square numbers in the total number of 5p coins.

Let the number of 5p coins be 'n' and let the number of heads showing at the end of the game be 'h'.
As the increase (75%) and decrease (50%) of n are exact, n must be divisible by 4, and must be less than 200 (the total value is less than £10).
As the increase (40%) and decrease (40%) of h are exact, h must be divisible by 5.
The minimum that n can be to show h number of heads is h^2, so h must be less than 15. Therefore, h is 5 or 10.

Possible solutions are:

n	h	75% increase	h	50% decrease	h
28	5	49	7	14	3
32	5	56	7	16	4
100	10	175	13	50	7
104	10	182	13	52	7
108	10	189	13	54	7
112	10	196	14	56	7
116	10	203	14	58	7
120	10	210	14	60	7

Only the first of these gives a 40% increase in h and a 40% decrease in h for the changes in n.

So n must be 28 and the value of the savings is £1.40.

Answer: 40699604

Write the palindrome product $P = L \times S$, where L and S denote the larger and the smaller numbers respectively. If d is the number of digits in S, we write $S = s_d \ldots s_1$ and $L = l_{(9-d)} \ldots l_1$.

Write $Q = L + S + 100$, another palindrome. We examine the cases d = 1 to 4 separately.

Case d = 1: Of Q's eight digits, at least five are different from each other and so Q is not a palindrome.

Case d = 2: Since we are given $s_1 = 3$ and $l_1 \times 3$ ends in 4, we know that $l_1 = 8$; thus $q_1 = 1$. Since Q is a palindrome and has 7 digits, $l_7 = q_7 = 1$. If $s_2 > 2$, $P > 1200000 \times 43 > 40699604$ so is not the smallest answer. Hence S = 23 and since $1700000 \times 23 < 40000004$, l_6 can't be 4, 5, or 6. As 8 is not available and $1900000 \times 23 = 43700000$ is greater than my son's answer 43122134, we have $l_6 = q_6 = 7$ and therefore $l_2 = 4$. As Q is a palindrome, the remaining digits 5, 6 and 9 force L = 1769548 and hence P = 40699604.

Case d = 3: If there is a smaller P, it must therefore satisfy $400000004 < P < 40699604$. $L = l_5 l_4 l_3 l_2 8$ and $S = s_3 s_2 3$. As in the previous case $s_3 = 2$ and $l_5 = 4$, 5 or 6 since $l_5 = 7$ makes the product P too big. For $l_5 = 4$, Q is not a palindrome except for the case Q = 149568 + 273 + 100, but then P = 149568 x 273 = 40832064 is not palindromic. If $l_5 = 5$ or 6, Q cannot be a palindrome; for example, if L starts with 15, then Q starts with 15 unless $l_4 = 9$, $l_5 = 7$ and $l_2 + s_2 = 16$, but in this case, $l_2 + s_2 = 8 + 8$, or 7 + 9 are not possible because 7, 8 and 9 have already been used. The equation $l_2 + s_2 = 15$ is similarly ruled out.

Case d = 4: As 17008 x 2403 = 40870224 > 40699604, we know that $l_4 = 4$, 5 or 6. Suppose $l_4 = 4$. Since $l_3 + s_3 > 10$, it implies that $q_4 = q_2 = 7$, and hence that $l_2 + s_2 = 16$. Consequently $(l_2, s_2) = (9, 7)$ or (7, 9) and $(l_3, s_3) = (6, 5)$ or (5, 6). However, none of the four products arising from these values (e.g. 14578 x 2693) is a palindrome. Similarly, if $l_4 = 5$ or 6, then $q_2 = 8$ or 9, but there are no viable values for l_2 and s_2.

94 PRIMARY ROAD NETWORK

Answer: 607, 829 and 1435

Let A, R and T be the numbers of areas, roads and towns respectively. I will demonstrate that A+T = R+1.

Any network can be built from scratch, adding a road at a time. If we start with a solitary town, A=R=0 and T=1, so the formula works. Adding a road to the existing network (R increases by 1) either (a) adds a new town without creating any new areas, so T increases by 1, A is unchanged and the formula still works or (b) connects to an existing town, adding a new area, so A increases by 1, T is unchanged and the formula still works. This is Euler's formula for maps.

A, R and T use ten different digits between them. We have three possibilities for A+T = R+1:

(1) a+bcde = fghij + 1. This doesn't work since b=c=d=9

(2) ab+cdef = ghij + 1. In this case, we must have d=9 (otherwise c=g). Since ab and cdef are prime, b and f must come from 1, 3 and 7.
(i) If b,f = 1,3 then j=3, but we've already used 3
(ii) If b,f = 1,7 then j=7, but we've already used 7
(iii) If b,f = 3,7 then j=9, but we've already used 9

(3) abc+def = ghij+1. In this case, we must have g=1. Since abc and def are prime, c and f must come from 3, 7 and 9.
(i) If c,f = 3,7 then j=9
(ii) If c,f = 3,9 then j=1, but we've already used 1
(iii) If c,f = 7,9 then j=5

Now A+R+T must have a digital root (remainder on divisibility by 9) of 9 since it uses all ten digits, and A-R+T = 1, so R must have a digital root of 4. Taking the remaining two possibilities above, we have:

(i) We are looking for 1hi9 with a digital root of 4 where h and i don't include 3 or 7. The only possibility is 1489 (1849 is too large). The remaining digits are 0, 2, 5 and 6, so abc+def can't make 1490.

(iii) We are looking for 1hi5 with a digital root of 4 where h and i don't include 7 or 9. The only possibilities are 1345 and 1435 and the remaining digits are 0, 2, 6 and 8. abc+def can't make 1346, but it can make 1436. a and d are 6 and 8 in some order, and b and e are 0 and 2 in some order. The one possibility that gives prime numbers for abc and def is 607 + 829 = 1436, since 627, 807 and 609 are divisible by 3.

Answer: 8cm x 12cm, 102cm

Each piece of wire joining two nails is the hypotenuse of a right-angled triangle with width a, height b and hypotenuse c that conform to $a^2 + b^2 = c^2$. The width and height are always whole numbers of centimetres, because adjacent nails are 1cm apart, but the hypotenuse is an irrational number unless (a, b, c) is a Pythagorean triple, in which case it is a whole number. The total length of wire is a whole number so all of the hypotenuse lengths belong to Pythagorean triples. We will only need the smallest ones of these, namely (3-4-5), (6-8-10) and (5-12-13), since the width plus height cannot exceed 20cm (half the perimeter).

With a pack of 40 nails it is possible to form many different rectangles but one dimension must be at least 3cm and the other at least 4cm in order to accommodate the smallest triple. The rectangles of size 3 x 4 up to 10 x 10 are evaluated using the process described below for the 7 x 12 rectangle. It is only necessary to consider starting in one quadrant because a rectangle is symmetric, so in the case of the 7 x 12 rectangle described below it is only necessary to consider starting at 1, 2, 3, 4, 5, 6, 7, 36, 37 or 38.

Nail 1 can only be connected to 18 (5-12-13) for a length of 13. Nails 2, 3, 6 and 7 cannot be connected to other nails. Nail 4 can only be connected to 35 (3-4-5) then 35 to 28 (3-4-5), 28 to 14 (6-8-10), 14 to 33 (5-12-13), 33 to 9 (6-8-10), 9 to 16 (3-4-5) and 16 to 23 (3-4-5) for a length of 53. Nail 5 can be connected to 19 (6-8-10) or 36 (3-4-5) but the fainter wires have a length of 10, whereas the thicker wires have a length of 43. Following this process for nails 36, 37 and 38 shows that the maximum possible length of wire in this rectangle is 53cm.

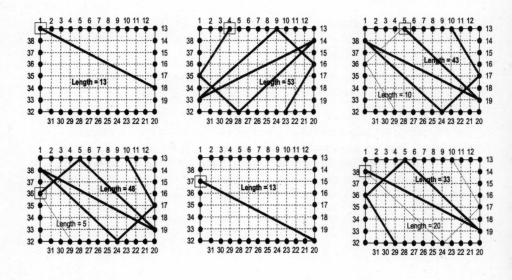

Following the above process for all the other rectangles shows that the longest possible wire is 102cm. This can be achieved in an 8 x 12 rectangle by starting at nail 14, 20, 34 or 40. The solution is shown in the diagram below.

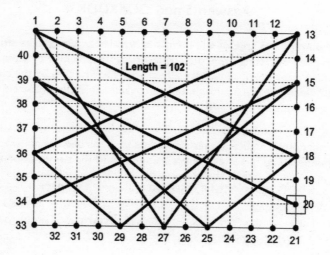

Answer: 5 and OOKKDDK

Progressively examine the 'tree' of all potential sequences of patterns as strings of Ds, Ks & Os, each pattern starting at the first 'rest' left by the previous patterns. This is straightforward, except when a K pattern appears for the first time in the composition; then, for that branch, the possibilities must be considered for the middle K beat not given in the conditions.

In the working below, the beats are denoted by lower case letters, and 'rests' by '/'. Subscript numbers for K indicate when its middle beat has been established on that branch. Bold underlining indicates the latest additions to the composition.

Patterns	Beats so far in the composition	Comment
D	**d**/**/d**/**ddd**/**dd**	only K could follow
D**K**	d**k**/**d**/ddd**k**dd	nothing can start at beat 3, so K must have middle beat 2
DK$_2$	dk**k**d/dddkdd	nothing can follow
K	**k**/////**k**	K could have middle beat 2; or else another K, or O, can follow
K$_2$	k**k**/////k	only K can follow
K$_2$**K$_2$**	kk**kk**///k/**k**	only K can follow
K$_2$K$_2$**K$_2$**	kkkk**kk**/k/k/**k**	nothing can follow
K**K**	k**k**/////k**k**	K could have middle beat 3; or else another K can follow
K$_3$K$_3$	kk**kk**///kk	only K can follow
K$_3$K$_3$**K$_3$**	kkkk**k**/**kkk**//**k**	nothing can follow
KK**K**	kk**k**////kk**k**	K could have middle beat 4 or 5; or else another K can follow
K$_4$K$_4$K$_4$	kkk**kkk**/kkk	nothing can follow
K$_5$K$_5$K$_5$	kkk/**kkk**kkk	nothing can follow
KKK**K**	kkk**k**///kkk**k**	not enough spaces for middle beat of four Ks; so fails
K**O**	k**oo**//**ook**//**o**	nothing can start at beat 4, so K must have middle beat 4
K$_4$O	koo**k**/ook//o	nothing can follow
O	**oo**//**oo**///**o**	only O can follow
O**O**	oo**oooooo**/o/**o**	only K can follow
OO**K**	oooooooo**ko**/o///**k**	K could have middle beat 3, or else K can follow
OOK$_3$	oooooooooko**ko**///k	K or O can follow
OOK$_3$**K$_3$**	oooooooookoko**k**/**kk**///**k**	nothing can follow
OOK$_3$**O**	oooooooookoko**oo**/**koo**///**o**	nothing can follow
OOK**K**	oooooooooko**ko**///k/**k**	K's middle beat could be 5 or 7
OOK$_5$K$_5$	oooooooookoko**k**/**kk**/k	only D can follow
OOK$_5$K$_5$**D**	oooooooookokok**dkk**d**kddd**/**dd**	D or K can follow
OOK$_5$K$_5$D**D**	oooooooookokokdkkdkddd**dddd**/**ddd**/**dd**	only K can follow
OOK$_5$K$_5$DD**K$_5$**	oooooooookokokdkkdkdddddddd**kddd**k**ddk**	complete! – valid solution

$OOK_5K_5DK_5$	oooooooookokokdkkdkdddkdd/k//k	only D can follow
$OOK_5K_5DK_5\underline{\mathbf{D}}$	oooooooookokokdkkdkdddkddd**k/d**k**ddd/dd**	nothing can follow
OOK_7K_7	oooooooookoko//**kk**k	only K can follow
$OOK_7K_7\underline{\mathbf{K_7}}$	oooooooookoko**k**/kkkk**kk**	nothing can follow

So there is a unique solution, as shown, with 33 quaver beats in the composition. If the seven instruments are labelled in their order of appearance, the 33 beats are played by the instruments as follows.

112211223142354354555655676667667

Answer: 1331, 2401, 4096, 4913, 5832, 6561

4-digit square of square options are:- $36^2=1296$; $49^2=2401$; $64^2=4096$; $81^2=6561$

4-digit cube options are:-

10	11	12	13	14	15	17	18	19	20	21
1000	1331	1728	2197	2744	3375	4913	5832	6859	8000	9261

The only options to isolate two sets of three values with lowest not square are:
(S is square of square, s is square of non-square, c is cube of non-square)

cSc	csS	three S values overall must contain numerals 0, 1, 2, 4, 6, 9
sss	Ssc	and possibly 5 from above options
ScS	csS	so unused numeral is 3, [5], 7 or 8

1st scheme:
lowest c<1296 (1000), unused numeral unknown - no other cells deducible with certainty e.g.

1000	1296	1728			**1000**	2401	3375	
1764	1849	2025	no '3'		3600	3721	3969	no '8'
2401	2744	4096			4096	4913	6561	

lowest c >1296, S values forced to 2401, 4096 and 6561 (numeral 3, 7 or 8 unused)
So next c >2401 <4096 =2744 or 3375 (Watson can't be sure, but numeral 7 used)
Highest c >4096 <6561 (4913 or 5832 - numeral 3 used - unused numeral=8, highest c=4913)

[1331	2401	2744]	[1331	2401	3375]	[2197	2401	2744]	[2197	2401	3375]
[2916	3025	3136]	[3600	3721	3969]	[2916	3025	3136]	[3600	3721	3969]
[4096	4913	6561]	[4096	4913	6561]	[4096	4913	6561]	[4096	4913	6561]

only four other cells known for sure (underlined) — not a majority

2nd scheme:
lowest c<1296 (1000), unused numeral unknown - no other cells deducible with certainty e.g.

1000	1024	1296			**1000**	1089	2401	
2401	2500	2744	no '8'		4096	4900	4913	no '7'
3375	3600	4096			5832	6400	6561	

lowest c >1296, S values forced to 2401, 4096 and 6561, but now two other c values >4096 <6561 must be 4913 and 5832 (so missing numeral is 7), so majority of other cells known
Lowest c >1296 <2401 =1331 (not 1728 or 2197 since 7 is missing)
So Watson knows 1331, 2401, 4096, 4913, 5832 and 6561

1331	[2025]	2401	
4096	[4900]	4913	Watson told 1331 and now sure of five others (majority - underlined)
5832	[6400]	6561	[squares of non-squares chosen from several possible]

Answer: 19, 31, 29, 16, 25, 23, 28 and 27

It is simpler to look at the pairs of names that have no letter in common: in the diagram such pairs are joined by a line.

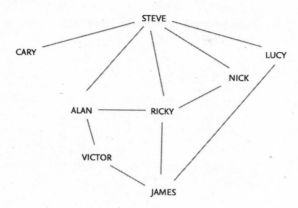

The possible three-figure squares are listed below, and for each (in brackets) are those squares having no digit in common with it:
169/196/961 (324, 784)   256/625 (784, 841)   289 (361, 576)   324 (169, 196, 576, 961)
361 (289, 529, 729, 784)   529 (361, 784, 841)   576 (289, 324, 841)   729 (361, 841)
784 (169, 196, 256, 361, 529, 625, 961)   841 (256, 529, 576, 625, 729)

Steve is joined to five others so his square must be 784 or 841. But in the case of 841 none of the five joined to Steve would be joined to each other. **So Steve≡784.**

Alan and Ricky are amongst those joined to Steve and they are joined to each other. So they must be 361 and 529 in some order. James is joined to four others including 361 or 529 and so **James≡841**. Therefore James is joined to 529 but not 361, making **Alan≡361** and **Ricky≡529**.

Lucy and Nick are joined to Steve and James and so (with Lucy being younger) we must have **Lucy≡256** and **Nick≡625**.

Victor is joined to James (841) and Alan (361) so **Victor≡729**.

Finally, Cary is joined to 784 but we've already used 256, 361, 529 and 625. So Cary is one of 169/196/961. Since Lucy is the youngest we have **Cary≡961**.

Therefore their ages are:
Alan 19, Cary 31, James 29, Lucy 16, Nick 25, Ricky 23, Steve 28 and Victor 27.

Answer: 1901, 1907, 1931, 1949, 1979, 1993, 1997 and 1999

The years in the twentieth century that have prime numbers were 1901, 1907, 1913, 1931, 1933, 1949, 1951, 1973, 1979, 1987, 1993, 1997 and 1999. We are told to remove 1973 from our consideration. The following table shows the intervals between any two of the other prime years which are entirely within the twentieth century, and which do not exceed 30.

	07	13	31	33	49	51	79	87	93	97	99
01	6	12	30								
07		6	24	26							
13			18	20							
31				2	18	20					
33					16	18					
49						2	30				
51							28				
79								8	14	18	20
87									6	10	12
93										4	6
97											2

The only term here beginning in 1951 contains 27 or 28 years, neither of which numbers is prime, so there was no change of prime minister in 1951. We show the years of service which correspond to the prime numbers 2, 3, 5, 7, 11, 13, 17, 19, 23 and 29, as follows:

2: 1931-33; 1997-99. 3: 1993-97. 5: 1901-07; 1907-13; 1987-93; 1993-99. 7: 1979-87. 11: 1901-13; 1987-99. 13: 1979-93. 17: 1913-31; 1931-49; 1979-97. 19: 1913-33; 1979-99. 23: 1907-31. 29: 1901-31; 1949-79.

As there were 9 Prime Ministers during the century altogether, we need to find 7 different prime numbers for the lengths of terms of office in the period between the first and last years in the century when there were changes of prime minister. We shall consider later the terms of office overlapping the ends of the century.

First, the only possible term of office ending in 1987 is 1979-87, so let us try to build a chain including this term, but not involving a change of Prime Minister in 1973. The only such chain is 1907-31-49-79-87-93-97-99, which includes terms of office of 23, 17, 29, 7, 5, 3 and 2 years respectively, and then the term ending in 1907 would have started in the prime year 1889, so it lasted 17 years, which length of tenure is repeated later in the chain. No chain including a change in 1987 is therefore possible, which means that no possible chain includes a term of office of 7 years, the figure 8 appearing only once in the table above.

Secondly, suppose that all the prime ministers had at least 3 completed years of service. The smallest sum of 7 such prime numbers, not including 7, is 3+5+11+13+17+19+23 = 91, to which sum we need to add 7 for uncompleted years of service, so the total number of years between first and last changes of Prime Minister is 98, indicating that those years were 1901 and 1999. However, we need a period of at least 29 years between 1949 and 1979 (as 1951 and 1973 are out), so a set of prime numbers that are all odd is not possible.

Thirdly, therefore, we look for a set of 7 prime numbers, all less than 30, including 2 and 29 but not 7, and adding 6 to their sum for the uncompleted years of service, so that the new total equals the difference in years between the first and last changes of Prime Minister. We find that there is one such set, namely 2,3,5,13, 17, 23 and 29, spanning the period from 1901 to 1999, and giving rise to just one possible set of periods of office:
1901 – 1907 – 1931 – 1949 – 1979 – 1993 – 1997 – 1999.

For the terms of office of the two Prime Ministers who overlapped the ends of the century, there could be one who took office in 1889 and served for 11 completed years up to 1901, and one who took office in 1999 and served for 7 years up to 2006 or 2007, which concluding year does not need to be a prime year.

100 CHECKMATE

Answer: 10

As the grand total of matches has decreased it seems much more likely that the number in each section has decreased.

As the number of players has only decreased by two it is most likely that the number of sections has increased.

Let the number of sections be a.
Let the number in each section be n.

$$na = (n - 1)(a + 1) + 2$$
$$a = n + 1$$

Considering the number of matches played, each of n players plays against n-1 others, resulting in n(n-1)/2 games, so

$$\frac{an(n - 1)}{2} = \frac{(a + 1)(n - 1)(n - 2)}{2} + 63$$

$$n(n + 1)(n - 1) = (n + 2)(n - 2)(n - 1) + 126$$

$$n^2 + 3n - 130 = 0$$

$$(n + 13)(n - 10) = 0$$

$$n = 10$$

For completeness we need to consider other possibilities.
Suppose the number in each section has decreased and so has the number of sections-

$$na = (n - 1)(a - 1) + 2$$

$$n + a = 3 \qquad \text{This is clearly impossible.}$$

Suppose the number of sections has decreased but the number in each section has increased-

$$na = (n + 1)(a - 1) + 2$$
$$a = n - 1$$

$$n(n - 1)^2 = n(n + 1)(n - 2) + 126$$

$$n^2 - 3n + 126 = 0 \qquad \text{There are no integral solutions to this equation.}$$

There are therefore 10 players in each section.

QUICK ANSWERS

Quick Answers

1.   31589712
2.   47
3.   100010101110
4.   5159 and 5951
5.   45, 91 & 990
6.   +1274, +1289
7.   729
8.   1, 2, 5, 8 and 9
9.   1,2,3,4,5,6,7
10.   4 yards
11.   32775
12.   Catter, Eighties, Forkpoynt and Hegroom.
13.   17 in.
14.   381654729
15.   630mm and 210mm
16.   1, 5 & 22
17.   101 and 146
18.   Ace of clubs and nine of spades
19.   22 pupils, 1052 ohms
20.   7, 7 & 4 blocks
21.   315
22.   11
23.   1, 2, 3 and 6
24.   8 & 5 centimetres, and 16; 7; 4 & 17; 10
25.   37
26.   41
27.   1, 2, 1, 2, 2 & 1 pesos
28.   2, 6, 7, 9, 18 and 42
29.   67, 31 and 53
30.   10
31.   Rathripe, Transponder, Vexatious and Wergild
32.   17
33.   7753222 and 7553332
34.   26 feet
35.   5ft by 7ft
36.   109
37.   397cm
38.   5,6,9,6,5,6
39.   583
40.   Spanish, Gloucester
41.   49, 53, 216, 780
42.   13, 33, 37 and 63
43.   13/75, 7/15, 3/5, 19/25
44.   Carlton, Grafton, Barton
45.   1188
46.   306, 918 and 2754
47.   2, 13 & 17
48.   16
49.   15
50.   41

51.	65km
52.	£6,660
53.	29 February 1796
54.	EIGHT SNAP THREE KING ACE SHOUT
55.	500 and 165
56.	8
57.	KNO
58.	£32
59.	1 2 2 3 3 4 4 5
60.	4, 30, 40 and 74
61.	1815
62.	7
63.	15
64.	127 feet
65.	18, 59
66.	5, 37 and 97
67.	8/15
68.	15625, 34969, 62001 and 91809
69.	311
70.	3, 17 & 35 and 5, 7 & 9
71.	826, 574 and 536
72.	11 and 26th
73.	1771
74.	13, 17 and 19 litres
75.	36 red, 67 yellow, 55 blue and 42 green
76.	96 72 64 54 48 32 16
77.	A33, B35, C36, D36
78.	1027
79.	274
80.	198752346
81.	BD 51 HLX
82.	105
83.	Knife, motor bike
84.	420cm
85.	176
86.	159, 267 and 348°
87.	5, 3, 6, 12 and 7
88.	Ann, Ben, Dave, Celia, Ben
89.	19 tiers; 12 and 5 years.
90.	171
91.	1804
92.	£1.40
93.	40699604
94.	607, 829 and 1435
95.	8cm x 12cm, 102cm
96.	5 and OOKKDDK
97.	1331, 2401, 4096, 4913, 5832, 6561
98.	19, 31, 29, 16, 25, 23, 28 and 27
99.	1901, 1907, 1931, 1949, 1979, 1993, 1997 and 1999
100.	10